MONOGRAPHS ON CRYOGENICS

in collaboration with
the British Cryogenics Council

Editor: R. G. Scurlock
University of Southampton

MONOGRAPHS ON CRYOGENICS

---

2. Martin N. Wilson: *Superconducting magnets*
3. Randall F. Barron: *Cryogenic systems*, second edition
4. René A. Haefer: *Cryopumping theory and practice*
5. W. H. Isalski: *Separation of gases*
6. Graham Walker: *Miniature refrigerators for cryogenic sensors and cold electronics*
7. V. Kovachev: *Energy dissipation in superconducting materials*
8. Ralph G. Scurlock (ed.): *Origins and history of cryogenics*
9. G. Walker and E. R. Bingham: *Low-capacity cryogenic refrigeration*

Low-Capacity Cryogenic Refrigeration

G. WALKER AND E. R. BINGHAM
Department of Mechanical Engineering
The University of Calgary, Alberta, Canada

CLARENDON PRESS · OXFORD
1994

Oxford University Press, Walton Street, Oxford OX2 6DP
Oxford New York Toronto
Delhi Bombay Calcutta Madras Karachi
Kuala Lumpur Singapore Hong Kong Tokyo
Nairobi Dar es Salaam Cape Town
Melbourne Auckland Madrid
and associated companies in
Berlin Ibadan

Oxford is a trade mark of Oxford University Press

Published in the United States
by Oxford University Press Inc., New York

A catalogue record for this book is available from the British Library

Library of Congress Cataloging in Publication Data
Walker, G. (Graham), 1930–
Low-capacity cryogenic refrigeration / G. Walker and E. R. Bingham.
(Monographs on cryogenics ; 9)
Includes bibliographical references (p. –) and index.
1. Low temperature engineering. I. Bingham, E. R. (Edward Reid)
II. Title. III. Series.
TP482.W35 1994 621.5′9–dc20 93–46643
ISBN 0 19 851760 2

Typeset by Apex Products, Singapore
Printed in Great Britain by
Bookcraft (Bath) Ltd, Midsomer Norton, Bath

Micro-forward looking infrared unit with miniature integrated dewar/cooler assembly (courtesy Ricor Ltd., Israel).

FOREWORD

by

Klaus D. Timmerhaus
University of Colorado

The past 40 years have seen many changes and improvements in low-temperature technology and most definitely in cryogenic refrigeration systems and components. During that time the demands for increasing numbers and types of miniature cryogenic coolers have been spurred not only by ambitious space missions but also by the need for helium refrigeration systems for use in a wide variety of applications utilizing superconducting magnets. The principal requirements for these applications include adequate refrigeration at specified temperatures with low power input, long lifetime, reliable and maintenance-free operation with minimum vibration and noise, compactness, and light weight. The need for reliability has provided the impetus for the development of several new low-temperature cooling devices, commonly referred to as cryocoolers, and the continued refinement of various components associated with such devices.

Cryocoolers are generally defined as refrigeration systems that are capable of achieving temperatures below 120 K for an extended period of time. Gas cycle cryocoolers are further classified by the method of heat transfer employed in attaining these low temperatures, that is, whether regenerative or recuperative methods of heat exchange are used. Every cryocooler requires essentially three distinct parts: a warm end, a heat exchange region, and a cold end. The warm end usually involves a compressor utilizing either external electrical power or heat energy to accomplish the compression. Refrigeration takes place at the cold end and normally involves expansion devices such as a throttling valve, expander, displacer, pulse tube, or acoustic resonator.

Because of the many advances in cryocoolers that had occurred during the 1970s and early 1980s, it was most fortunate that the general editors of the Plenum Publishing Corporation, New York, were able to interest and persuade Professor Graham Walker, a world-wide authority on cryocooler technology and operation, to carefully review this progress and summarize it along with his own highly regarded research endeavors in a very successful, as well as popular, two-volume monograph entitled *Cryocoolers: Parts I and II*. The first volume is an excellent summary of the fundamentals that serve this entire area of cryogenics. Included in the discussion are detailed analyses of the Stirling, Vuilleumier, Gifford–McMahon, Joule–Thomson, and Claude cryocoolers. The section on Stirling cryocoolers, in particular, is one of the most definitive expositions in the open literature on this highly successful type of cryocooler. Citation Index records shows that this section of the monograph has been referenced more often and by more authors than any other cryogenic paper published in the mid-1980s. The second volume places an emphasis on the hardware associated with these cryocoolers, notes some key design aspects involved in the integration of the hardware, and outlines potential operating problems and their solutions. This volume also considers alternative cooling systems, including various ultra-low-temperature cooling devices. It concludes with authoritative guest reviews of cryocooler development and progress in the former USSR and Japan. *Cryocoolers: Parts I and II* has served the cryogenic community admirably for the past decade and will provide those entering the field with an exceptional introduction to the world of cryocoolers.

As in every active field of science and engineering, however, progress is inevitable. It comes as no surprise that this is also the situation in the rapidly changing technology associated with cryocoolers. The latter have become increasingly application-specific with a sophistication unknown even in the early 1980s. Thus an up-to-date volume about low-capacity cryogenic refrigeration is most timely.

A preliminary review of the manuscript for this new monograph shows that Professor Walker has very succinctly summarized the fundamentals of cryocoolers that he had developed in his earlier monograph and then has carefully guided the reader through the more recent advances in this fascinating technology. Specifically, after reviewing the many applications, the monograph summarizes the concepts involved with open- and closed-cycle cryocoolers. The latter examines systems utilizing the latest advances in regenerative and recuperative heat exchangers.

All users of cryocooler systems are indebted to Professor Walker for his continued willingness to share his expertise of cryocoolers with them. In the monographs that he has developed he has blended his expertise with the latest advances in this special area and has provided a unique service to all in the cryogenic community. All of us thank him for this dedicated service and wish him well in any future endeavours.

PREFACE

Nearly a decade and a half has elapsed since the draft was completed of an earlier work, *Cryocoolers*†. The book was well received and appears to have been a useful and timely contribution. It is gratifying to see the familiar blue covers on the bookshelves of many colleagues visited from time to time and the frequency of citation in many papers and books. Nevertheless, the lapse of 15 years in a field so dynamic is reason enough to revisit.

Moreover, having used the book for short courses at various places, principally the University of California, Los Angeles, we have come to recognize the need for a different book; one that is more compact, lower in cost, yet which embraces more topics and is reasonably up to date.

This is our attempt to fill that need. It is intended as an introduction both to those entering the field and to others whose principal interest is not in cryogenics or cryocoolers but who wish to gain some appreciation and understanding of the problems and opportunities therein.

Low-capacity cryogenic refrigeration is presently used for cooling infrared detectors in night vision and missile guidance systems and in a wide variety of commercial, industrial, medical, scientific, and educational applications.

Use on a broader front is anticipated with the commercial development of 'cold electronics' including the introduction of high-temperature (80 K) superconducting electronic devices.

There is a growing conviction that the use of cold electronic devices, both superconducting or semiconducting, will not become widespread until small, compact cryogenic refrigerators, cryocoolers, are available which are reliable, work for long periods without maintenance, and can be operated simply by closing a switch.

The demand for improved reliability and efficiency of the small cryogenic refrigerators used by the military for cooling infrared detectors has led to many significant developments in the last ten years. Some of the new developments in reliability have resulted from the space program where lifetimes of 5–10 years are usually needed. This book outlines the various approaches taken to improve reliability and efficiency.

Recognition of the critical significance of cryocooler development to future use of superconducting and semiconducting electronic devices, along with other cryogenic sensors, led to a meeting of invited specialists

† Walker, G. (1983) *Cryocoolers*, 2 vols, Plenum, New York.

at the US National Bureau of Standards in 1978 and subsequently to the establishment of biennial 'Cryocooler Conferences' with unrestricted participation and published proceedings. Seven conferences have now been held, in 1980, 1982, 1984, 1986, 1988, 1990, and 1992. This book is, in some ways perhaps, a distillation and digest of these conferences with sufficient emphasis on fundamental aspects for those unfamiliar with the field to grasp the significance of a topic and its relation to competing systems. Reference is also made to other conferences, books, surveys, and other works that have significant relevance.

We have dedicated this volume to Klaus D. Timmerhaus, former Associate Dean of Engineering at the University of Colorado, Boulder, and now President's Teaching Scholar there. In addition to his many other activities Timmerhaus assumed the burden of editing the proceedings of the Cryogenic Engineering Conference. He did this not once, but more than twenty times. As contributors to the conference well know, Timmerhaus edited with scrupulous, meticulous, and insistent attention to detail. The magnificent result of this sustained and devoted effort is the collected proceedings of the Cryogenic Engineering Conference marching across one's bookshelf in a series of volumes as uniform and precise as Grenadier Guards. They are the encyclopaedic reference to cryogenics. Much credit for their creation belongs to Timmerhaus. Happily his editorial successors have maintained the same rigid standards that gives *Advances in Cryogenic Engineering* the value that it has.

We have many to thank for their help and assistance in the preparation of this book. We have tried throughout to acknowledge our sources of information, photographs, and the like. We apologize in advance for any omission or incorrect attributions, made inadvertently.

We particularly appreciate the contribution of Chapter 7, 'Mechanical cryo-refrigerators for spacecraft', by Dr Otto Ledford, Jr. This informative review of an important field of application of cryocoolers was first presented at the 1987 Cryogenic Engineering Conference. We are most grateful to Dr Ledford both for his kind permission to reprint the paper and for his revision of the material to include developments up to the end of 1992.

Karen Undseth typed the manuscript and prepared the index in her customary exemplary fashion for which efforts we are most grateful. Burt Unterberger and Ann Kiddle at the University of Calgary prepared the drawings and photographs contained therein.

We acknowledge with thanks the facilities and support extended to us at the University of Calgary, particularly the help and advice of our colleagues in the Department of Mechanical Engineering and the encouragement of Dr Graham Reader, Head of the Department.

The text was prepared in final form by Graham Walker in the course of a Sabbatical Leave Fellowship in the Winter Term of 1992 during which time the author was relieved of customary teaching and administrative duties, the burden of which fell upon the shoulders of his colleagues. For this relief he is most grateful.

Research on Stirling engines, including cryogenic refrigerators, is supported at the University of Calgary by the National Science and Engineering Research Council of Canada. We are most grateful for the continuing support from these various organizations.

Our greatest thanks are due our wives, Ann and Siobhan, for their unfailing support and encouragement.

Calgary
March 1994

G.W.
E.R.B.

This volume is dedicated to Klaus D. Timmerhaus.

CONTENTS

REVIEW

BUILDING THE BETTER CRYOCOOLER

Dr Ray Radebaugh
National Institute of Standards and Technology
M.S. 583.30
325 Broadway
Boulder, Co 80303

The following contribution by Dr Ray Radebaugh is reprinted (with permission) from the journal Superconductor Industry *(Fall 1992). It represents the state of development of cryocoolers as they are in 1993. It could well be the executive summary of this book. Dr. Radebaugh is distinguished scientist one of the leading contributors to the cryocooler field. For many years he has worked with Graham Walker and other colleagues to co-ordinate and present the annual cryocooler course at the University of California, Los Angeles. This book, as originally conceived, was to be the joint product of Walker and Radebaugh for use as the text on the annual course. In the event, pressure of work presented Radebaugh from contributing in a major way. This reprint serves well to display his encylopaedic understanding of the industry and convey his concepts for future developments.*

This article focuses on the present and future status of small cryocoolers that may be useful for cooling superconducting electronics or magnets no larger than those of magnetic resonance imaging (MRI) systems. In a few cases superconductors have found their way into a large market. The best example is superconducting magnets for MRI. Magnets for the superconducting super collider (SSC) and other accelerators use an enormous amount of superconducting wire and require large liquid helium plants for cooling. The costs of these large accelerator systems greatly restricts the number of such installations.

Thin-film superconducting electronic devices have the potential of being made relatively inexpensively and have a performance advantage over conventional electronic systems. The possible market size for superconducting electronics could be extremely large if one serious problem would simply disappear: the need to cool the devices to cryogenic temperatures.

The manufacturer of a superconducting electronic device often has a very difficult time breaking into the existing electronic market-place when a potential customer asks to see all the support equipment associated with some hand-held superconducting device. A room full of noisy refrigeration equipment requiring water cooling, a building modification to install 5 kW of three-phase 230 V power, with a $30,000 to $50,000 price tag—all to cool a $10,000 superconductive device—quickly dampens any initial interest in the superconducting device.

Significant improvements have occurred in each type during the last ten years. Unfortunately, cost, size, and reliability are still common complaints against most cryocoolers.

Types of cryocoolers

According to the second law of thermodynamics, the input power per watt of refrigeration must be at least 2.8 W to achieve 80 K and 74 W to achieve 4 K. Small cryocoolers usually operate at 1 per cent to 10 per cent of Carnot efficient, with the higher efficiencies appearing more recently. Figure 1 shows schematics of the five common types of cryocoolers used at present. They are divided into two types: recuperative and regenerative.

The recuperative cryocoolers are analogous to a DC electrical device in the sense that the refrigerant flows steadily in one direction. This one-directional flow is often an advantage because they can transport the refrigerant over fairly large distances to do spot cooling at several locations. The recuperative heat exchangers have two separate flow passages and the streams continuously exchange heat with each other. Such heat exchangers are relatively expensive to manufacture. The one-directional flow requires that the compressor has inlet and outlet valves that add complexity and failure points.

The Joule–Thompson (JT) cryocooler cycle is very similar to the vapour-compression cycle used in household refrigerators except for the use of a non-CFC refrigerant to reach cryogenic temperatures and the need for a very effective heat exchanger to span such a large temperature difference. In a domestic refrigerator, oil from the oil-lubricated compressor dissolves in the CFC refrigerants and remains in solution even at the cold end.

Since both the CFC and the oil would be frozen at cryogenic temperatures, a different refrigerant must be used in cryocoolers, and it must be oil free when it enters the main heat exchanger. Reliable oil-free compressors are generally not available at present, although current research may lead to such a device that would also be inexpensive to manufacture. Often oil-lubricated compressors are used with the addition of highly efficient oil-removal equipment in the high-pressure outlet stream.

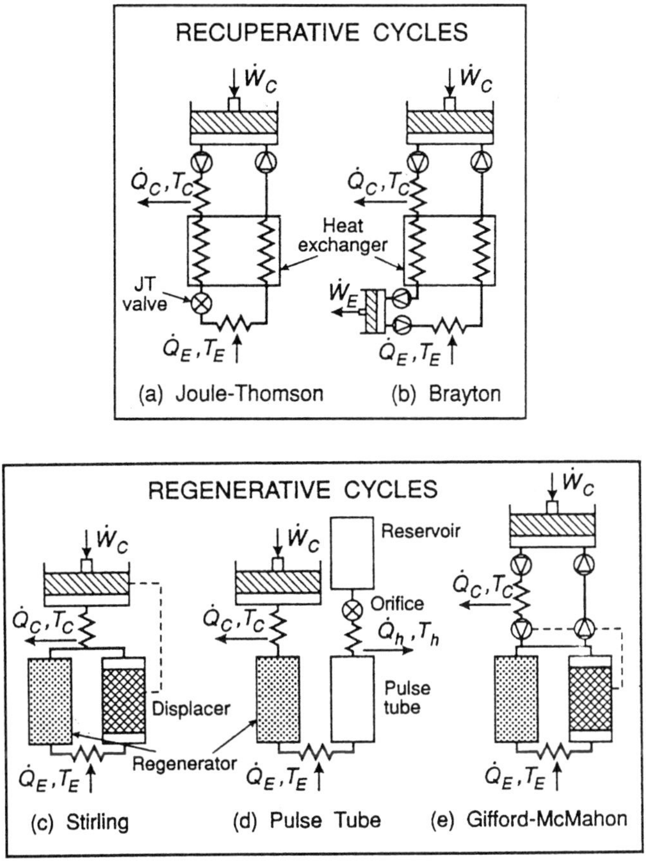

Fig. 1 Regenerative and recuperative cycles.

The irreversible expansion that occurs at the JT valve leads to cooling only for non-ideal gases below some inversion temperature. Nitrogen gas is typically used for refrigeration at 77 K, but the input pressure is usually 20 MPa (200 bar) to achieve reasonable efficiencies. Hydrogen gas, precooled by a nitrogen stage, is used for refrigeration at 20 K, and a helium stage is used to achieve 4 K. The three separate gases require three separate compressors, each with at least two stages of compression to achieve the high pressure ratios required. The overall system is very complex with many moving parts to wear out.

More often a 4 K JT system is precooled to 15–20 K with a regenerative refrigerator. A commercial version that produces 1 W at

4.2 K is available. It is precooled by a Gifford–McMahon refrigerator, requires about 4.5 kW of input power, and costs about $42,000.

Single-stage JT coolers that use nitrogen or argon with miniature finned-tube heat exchangers have been used in large quantities for rapid (a few seconds) cool-down of infrared sensors. These open systems use high-pressure gas from a small storage cylinder.

Another type of commercially available open system for 77 K operation uses etched glass plates for the heat exchanger and other flow channels. It provides about 0.35 W of refrigeration at 77 K (and above) and sells for about $1,000. Reliable compressors for these JT systems have generally not been available, although recent and significant progress has been made in such compressors.

The other common recuperative cryocooler is the Brayton cycle refrigerator. An ideal gas such as helium or a helium–neon mixture can be used on this cryocooler because of the reversible expansion that occurs in either the reciprocating or turbo-expanders. As a result, only one fluid is required for all temperatures and much lower pressure ratios are needed.

This cycle is commonly used in large liquefaction systems (with a final JT stage) and it has a high reliability due to the use of gas bearings on the turbo-expanders. This cycle is generally not practical or efficient for refrigeration powers less than 10 W at 80 K because of the machining problems encountered with such small turbo-expanders. As a result, its application to the cooling of superconducting electronics is rather limited.

Regenerative systems

Except for the Gifford–McMahon refrigerator, the compressor or pressure wave generator in the regenerative system has no inlet and outlet valves. As a result, it produces an oscillating pressure in the system, and void volumes must be minimized to prevent a reduction in the pressure amplitude.

Typical frequencies vary from about 2 Hz to 60 Hz. An oscillating displacer causes the gas to be compressed when it is at the warm end and to expand when it is at the cold end. In the pulse-tube refrigerator, the compressed, hot gas flows from the pulse tube through the warm heat exchanger and the orifice. The expanded cold gas in the pulse tube flows past the cold heat exchanger when gas from the reservoir returns to the pulse tube. These systems are analogous to AC electrical systems.

The primary heat exchanger is known as a regenerator or a regenerative heat exchanger. It consists of some form of porous material with high heat capacity, through which the working fluid flows in an oscillating manner. Heat is transferred from the fluid to a porous matrix (stacked screens or packed spheres) during the hot blow (fluid flowing from the warm end) and returned to the fluid from the matrix during the cold blow (fluid flowing from the cold end). Because of the single flow channel, regenerators are very simple to construct. The rapid decrease in heat capacity of most matrix materials at low temperatures causes a rapid decrease in regenerator performance below about 10–15 K.

As a result, all regenerative refrigerators are usually limited to temperatures above about 8–10 K. The working fluid in nearly all regenerative systems use helium gas. Temperatures down to about 50 K are usually achieved with single-stage cold heads, whereas two or more stages are used to achieve lower temperatures. From a thermodynamic standpoint, more stages lead to higher efficiency, but the additional manufacturing complexity must be considered in any practical device.

The Stirling refrigerator, which has the highest efficiency of all cryocoolers, is the oldest and most common of the regenerative systems. The Stirling cycle was invented for use as a power system in 1816 and first commercialized as a cryocooler in 1954.

The other regenerative refrigerators are variations of the Stirling cycle. Small, single-stage Stirling refrigerators are manufactured by the thousands by a dozen different manufacturers world-wide. They are the most common cryocooler used for cooling infrared detectors in tactical military applications. They are made in sizes from about 0.15 W to 1 W at 80 K. Largest dimensions are usually no more than 150 mm and masses are usually less than 3 kg. Power inputs are usually about 40 to 50 W per watt of refrigeration, which gives them a relative efficiency of 6–7 per cent of Carnot. As the refrigeration power increases, so does the efficiency. The highest efficiency achieved is 45 per cent of Carnot with 25 kW of refrigeration at 80 K.

The small tactical Stirling cryocoolers have lifetimes of about 2,000 to 4,000 hours and cost $5,000 to $10,000. The short lifetimes are a result of using dry-lubricated rubbing components in the compressor and the displacer. Most later-model Stirling coolers use linear motor drives rather than rotary drives with crankshafts to reduce the magnitude of rubbing forces and to achieve the lifetimes given above. A few manufacturers have recently announced the limited availability of larger Stirling cryocoolers with refrigeration powers in the 5–500 W range at 80 K.

Achieving longer lifetimes in these cryocoolers requires the elimination of rubbing components. At present, the most common technique for eliminating rubbing forces uses flexure diaphragm springs to radially centre the piston and the displacer in their respective cylinders.

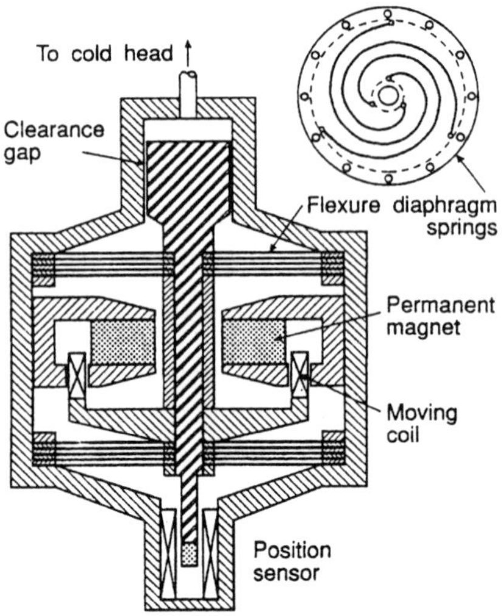

Fig. 2 Flexure diaphgram compressor.

The application of this concept to the Stirling cryocooler was first introduced by Gordon Davey at the University of Oxford in the early 1980s. (This concept in a compressor of a Stirling cryocooler is shown in Fig. 2.) There is a similar arrangement in the displacer that can be located up to a metre away through a small-diameter connecting tube. Vibration from the compressor or displacer is usually reduced by using two opposed units or a counterbalance mechanism. The seal between the piston and the cylinder is a clearance seal of about 15 to 2 μm. The amount of gas oscillating through this seal to the back side of the piston is only a few per cent of the gas moved by the piston in a half cycle.

Figure 2 shows a moving-coil linear motor for the drive mechanism, but a moving magnet or moving iron can also be used. The Oxford style of Stirling cryocooler is designed for satellite applications with lifetimes of five to ten years. Such a cooler was first placed into orbit aboard the NASA Upper Atmosphere Research Satellite (UARS) in September 1991 for cooling infrared sensors used in atmospheric studies. Though the experiment has failed, the cryocooler is still working as of this writing.

This type of Stirling cryocooler is now commercially available in a space-qualified unit with a power input of about 40 W and providing 0.35 W refrigeration at 65 K. It sells for about $500,000 with flight electronics and about half that with industrial electronics. Both the precise machining and alignment required for these Oxford Stirling cryocoolers contribute to their high cost. Other non-contacting mechanisms that have cost-reducing potential are gas or magnetic bearings in place of the flexure diaphragm springs.

Gifford–McMahon refrigerators were developed in the mid-1950s using the same type of cold head as the Stirling cryocooler. However, the pressure oscillation is generated by using valves to switch between the high- and low-pressure sides of an air conditioning compressor modified for use with helium gas. Oil in the high-pressure gas is removed by extensive filters and adsorbers before the gas enters the cold head. The use of valves to provide the pressure oscillation greatly reduces the system efficiency compared with the Stirling cryocooler, but it allows the use of inexpensive oil-lubricated compressors manufactured by the millions. These Gifford–McMahon refrigerators, now manufactured by the thousands for cryopumps, MRI shield cooling, ground-based satellite communications systems, and research applications, are available in both one- and two-stage units.

Single-stage units are useful down to about 40 K and two-stage units can achieve about 10 K. Refrigeration powers at 15 K can vary from about 0.5 W to about 7 W. Refrigeration powers at 77 K vary from about 15 W to 200 W and input powers range from about 1 kW to 8 kW. Prices are in the range of $10,000 to $25,000. System masses vary from about 60 kg to 200 kg. The adsorber needs replacement about once a year, the cold head needs the seals replaced about once every two to three years, and the compressors generally last about five years.

The pulse-tube refrigerator is a recent variation of the Stirling refrigerator in that the moving displacer is replaced by an orifice and reservoir volume. The original version of the pulse-tube refrigerator was developed in the mid-1960s, but a more powerful, orifice, version was not introduced until the early 1980s. The pressure oscillation is most commonly provided by a Stirling cycle compressor but a Gifford–McMahon compressor and valves are sometimes used with a sacrifice in efficiency.

In 1989 a thermoacoustic driver was used to drive a pulse-tube refrigerator in a joint project between NIST and Los Alamos National Laboratory. It achieved 90 K and became the first cryocooler with no moving parts. This heat-driven cryocooler is still at an early experimental stage and is not as efficient as mechanically driven pulse tubes. The overall system efficiency of orifice pulse tubes using Stirling-type compressors has, in the last few years, increased significantly to the point where they are now nearly as efficient as Stirling refrigerators (at least for sizes larger than a few watts at 80 K). The non-adiabatic behaviour within a very small pulse tube decreased the efficiency. A pulse tube of about 5 cm^3 can usually produce 5 W at 80 K when driven at a frequency of 40 Hz. Low temperatures of 30 to 40 K are possible with single-stage pulse tubes.

Little work on multiple-stage pulse tubes has been done at this time, but temperatures of 10 K should be possible. Advantages of the pulse tube are its increased reliability, reduced cost, and reduced vibration at the cold end because of the elimination of moving parts in the cold head. The disadvantage is its slightly lower efficiency at this time compared with Stirling cryocoolers. Pulse-tube refrigerators are not commercially available at present, but several manufacturers are experimenting with them.

Future trends

Progress toward developing a cryocooler designed specifically for cooling superconducting electronics has been slow due to the dependence on spin-offs from infrared cryocoolers. Those requirements are not always the same as those for superconducting electronics. As superconducting devices draw nearer to being manufactured, there will be greater emphasis on cryocoolers.

In Joule–Thompson cryocoolers significant improvements in efficiency have been made in the last few years by replacing pure nitrogen with a mixture of nitrogen, methane, ethane, and propane. Temperatures of 80 K can be achieved with four or five times the efficiency of nitrogen systems with lower pressures on the compressor output. An efficient, long-life compressor for the Joule–Thompson refrigerator is still needed, but some developments are in progress.

One development is in the use of sorption compressors. These compressors generate high pressure by desorbing gas at a high temperature, and low-pressure pumping occurs by using another sorbent bend cooled to ambient temperature. The only moving parts are a few check valves that switch once every few minutes.

There is now rapid progress in the development of long-life Stirling cryocoolers using flexure diaphragm springs. These may satisfy the short-term needs for satellite operation, but there is a need to reduce the cost of such devices. Some alternative non-contacting methods should be investigated further with the goal of reducing the cost of long-life Stirling refrigerators.

At some time in the future there will be more work on multiple-stage Stirling and pulse tube refrigerators to reach 15 K and to precool a JT stage to reach 4 K. Such systems could be much more efficient and smaller than the Gifford–McMahon/JT combination now used for 4 K refrigeration in small sizes.

With much more work on the development of high heat capacity materials for temperatures of 4 K to 20 K, Stirling, and even pulse-tube refrigerators, may eventually reach 4 K with good efficiencies. Because of its higher pressure ratio, the Gifford–McMahon refrigerator has recently achieved 4 K and liquefied helium in a few cases. Japanese researchers have been particularly successful in this area. Improved efficiencies can be expected in the future in these 4 K Gifford–McMahon refrigerators.

Since pulse-tube refrigerators have only been recently introduced, more improvements can be expected. Efficiencies will most likely be improved further as some of the loss mechanisms are better understood. They will begin to replace Stirling refrigerators in some applications, particularly where low cost, low vibration, and high reliability are very important.

Innovation is needed in the packaging of multiple-stage pulse tubes to keep them compact. Research on thermoacoustic drivers for pulse tubes is currently underway to increase the pressure ratio (currently 1.1 to 1.2) and efficiency (currently about 20 per cent of Carnot) to better compete with mechanical compressors Thermoacoustic drivers are particularly attractive where there is a natural head source that can be used directly.

The availability of reliable, low-cost, and efficient cryocoolers would have significant impact on the marketability of superconducting devices. Such cryocoolers do not exist at present, but progress toward this goal is slowly being made. A greater focus on this particular market is needed to speed up the progress and to have such cryocoolers available when the superconducting devices are ready for market.

1

INTRODUCTION, NOMENCLATURE, AND CLASSIFICATION

1.1 Introduction

Cryogenic temperatures are generally accepted to be temperatures less than 120 K†. Cryogenic refrigeration generally occurs within this temperature range, from low values less than 1 K to the upper bound and beyond. Refrigeration capacities range all the way from fractions of a watt to megawatts.

In this book we are concerned with low-capacity cryogenic refrigeration. However, there is no generally accepted classification so, for our purposes, we shall arbitrarily define low-capacity cryogenic refrigeration as 1 kW or less of refrigeration at a temperature of 80 K. This temperature is an important 'milestone', it being close to the temperature at which liquid nitrogen boils at atmospheric pressure. Liquid nitrogen is the cryogenic fluid most readily available in most places. It is cheaper than milk in most parts of North America, inert, and relatively safe to use. It is available in virtually unlimited quantities being produced very simply from atmospheric air whence it returns following the inevitable evaporation and does so with no impact on the environment.

It is always necessary to specify the reference temperature when discussing cryogenic refrigeration. A refrigerator capable of 10 watts of refrigeration at 4 K (the temperature of liquid helium) is very different from the refrigerator delivering 10 watts of refrigeration at 80 K.

There are many needs and requirements for cryogenic refrigeration. Some of the principal applications and their special or particular needs are discussed in Chapter 2. There are also many different ways to achieve cryogenic refrigeration. This book is primarily about some of the ways to do it, the fundamental technology, and the advantages and disadvantages of different methods.

† The Kelvin scale of temperature is widely used in cryogenics. 1 K (note the absence of any ° sign) is equal to 1 °C but commences at absolute zero. Other temperature scales are Rankine and Fahrenheit. Equal temperature increments are 1 K = 1 °C = 1.8 °R = 1.8 °F. On an absolute scale, 0 K = − 273 °C = 0 °R = − 460 °F (absolute zero); 273 K = 0 °C = 492 °R = 32 °F (the ice point).

1.2 Performance and physical parameters of low-capacity cryogenic refrigerators

In a review of cryogenic refrigerators for naval ship infrared detectors Voth (1974)† selected the following 28 performance and physical parameters of the different systems available. He assigned arbitrary 'points' to the various parameters, then he used the scheme to numerically rate 45 different cryogenic refrigerators of one and two watt capacity (at 77 K).

Parameter: Refrigeration temperature range (K)
Cycle
Working fluid
Lowest temperature attainable (K)
High pressure (MPa)
Low pressure (MPa)
Cooling time, no load (min.)
Cooling time, loaded (min.)
Temperature stability ($\pm$ K)
Expander rotational speed (H_3)
Compressor lubrication
Total power input (Watt)
Volts–phase–frequency
Heat rejected near detector (W)
Cooling medium
Ambient temperature requirements ($^\circ$C)
Required attitude
Cryostat volume (cm^3)
Compressor weight (kg)
Cryostat weight (gm)
Mean time before failure (MTBF) (hrs)
Mean time to repair (MTTR) (hrs)
Maintenance interval (hrs)
Vibration (frequency, amplitude, *g*-acceleration)
Cost 50–100 units ($)
Status

This study is now dated but the technique for numerical evaluation is still valid. Additional parameters may, of course, be added to the list and some removed. Depending on the application, a different 'points' allocation to that adopted by Voth might be used.

† References listed in full at the end of the chapter.

Cryocoolers are often rated by their *available refrigeration capacity* measured in watts. To be meaningful however, it is necessary to specify not only the refrigeration capacity but also the temperature at which the refrigeration is available, for example, 0.25 watt at 80 K.

Another important parameter is the *power input* or work required to achieve refrigeration.

The *coefficient of performance* of a refrigerator is defined as the ratio: COP = refrigeration capacity/power input; or heat lifted/work done. The ideal coefficient of performance is the Carnot value:

$$\mathrm{COP}_{(\mathrm{Carnot})} = \frac{T_E}{T_C - T_E} \tag{1.1}$$

where T_E is the minimum cycle temperature (usually the refrigeration temperature) and, T_C is the maximum cycle temperature (generally the atmosphere temperature).

The ratio of the actual coefficient of performance to the Carnot coefficient of performance is sometimes called the *efficiency*, an unfortunate term when applied to a refrigerator and one whose use is not to be encouraged. It is, however, a useful measure of the way an actual machine measures up to the thermodynamic ideal machine. Thus:

$$\text{Efficiency} = \frac{\text{Actual COP}}{\text{Carnot COP}}$$

$$= \frac{\text{Refrigerator capacity}}{\text{Power input}} \times \frac{T_C - T_E}{T_E}. \tag{1.2}$$

The efficiency of presently available cryocoolers ranges from a minimum of less than 1 per cent to a maximum near 50 per cent. Strobridge (1974) presented 'experience' charts for 140 different cryocoolers with the above efficiency variations. Similar, but more up to date, charts were given by Smith *et al.* (1984) and are reproduced in Chapter 4.

The small machines used for electronic applications have the lowest efficiencies. This is because practically all the refrigeration generated is consumed in cooling, and maintaining cold, the low-temperature region of the machine itself. The surplus or useful refrigeration available from these units is very small, for only fractions of a watt are required to cool the miniature cryogenic sensors or other silicon-based electronic devices for which they are used. Applications requiring a larger useful refrigeration load use systems bigger in size which tend to be more efficient. The highest efficiencies are found in large machines used for liquefiers, that is, tonnage oxygen plants, and range from 20 to 50 per cent of the Carnot value.

Other significant parameters for cryocoolers include the *total mass and volume* of the system and the *mass and volume of the cold region*. This latter is of particular importance in infrared (IR) missile guidance or night vision systems with cooled detectors and associated optical/electronic systems mounted in swivelling gimbals. A low mass and volume is necessary to facilitate a fast response with low inertia.

Cool-down time, the period necessary for the machine to achieve stable operation at the design condition following start-up, is important in some applications of miniature systems, usually weapons related. The combination of a quick cool-down and a slow warm-up is a particularly challenging requirement involving mutually opposed high and low 'thermal masses'. Cool-down is sometimes accelerated by operating at a high pressure or speed on start-up and later switching to the normal operating mode.

Mechanical vibration and *electromagnetic emissions* in the cold region are often important characteristics of cryocoolers. Some applications require the elimination of nearly all mechanical or electromagnetic noise. This is best approached by:

(1) physical separation of the cold region and the compressor unit where large power input and heat transfers take place;

(2) elimination of moving parts in the cold region;

(3) the use of plastic, ceramic, or other non-magnetic parts in the cold region.

Operating life is another important characteristic often quoted as the *mean time before failure* (MTBF) or the *mean time before maintenance* (MTBM). This is generally taken to mean the average time of operation before failure (or maintenance) of a number of identical ryocoolers operating under similar conditions. Reliable data for MTBF and MTBM based on the accumulated experience of users is very hard to come by for there appears to be no properly organized system for data collection, reduction, rationalization, and distribution.

Operating life is affected to a large extent by the nature of the system used. Moving parts, such as sliding seals, bearings, piston rings, and so on, wear and in doing so generate detritus with deleterious effects on valves, small passages, and close-tolerance clearance spaces. Sometimes questions of lubricated versus unlubricated parts involve balancing the increased wear against the possibilities of contamination by lubricant.

Reliability can often be improved by incorporating excess capacity so that operation is intermittent, or by the provision of redundant systems.

One solution implemented fairly recently is to eliminate wearing surfaces by the use of non-contacting seals and bearings. This results in systems where life and reliability depend mainly on the electronic units. These are typically much more amenable to predictable lifetimes based on piece part count and the environmental conditions.

There are few, if any, independent sources of data for the failure of identical cryocoolers used under identical conditions for the long periods of time necessary to generate meaningful MTBF or MTBM data. Moreover, systems evolve so that reliability data from a predecessor is not altogether applicable to an improved system. The lifetime of wearing components cannot yet be assessed theoretically and can be characterized only by adequate experimental testing.

The *shelf life* of a cryocooler is frequently important. This is the interval a unit may be required to maintain the capability for operation while standing unused or in occasional intermittent use. Many systems involve mechanical units containing high-pressure helium and are equipped with a variety of sealed flanges, shaft seals, screwed fittings, and so on. Some leakage of the working fluid is almost inevitable over the long term. Hermetic sealing by seal welding is the best approach to ensure a long shelf life.

In *split systems* where the cold region is separated from the power input unit the nature of the tubes or leads connecting the two elements is important. Some systems have flexible tubes, others are inflexible but may incorporate swivel joints. Sometimes the leads are cold, sometimes at ambient temperature, sometimes a single tube with a cyclic fluctuating pressure, sometimes twin tubes with a high-pressure feed line and a low-pressure return. All these affect the insulation required, the forces and bending moments imposed on gimballed units, the transmission of mechanical vibration, thermal conduction, and pressure attenuation or phasing effects.

Cost, of course, is always important, more so in some cases than others. Separate consideration of capital cost and operating cost including scheduled or unscheduled maintenance and replacement is necessary. Sometimes the lack of opportunity or even the possibility for maintenance (space systems) justifies an increase in the capital cost by the provision of redundant systems or the use of a large, lightly loaded, slow-running unit.

The power supplied to drive the cryocooler is eventually degraded to heat and must be rejected from the system. Thus the *cooling system* to carry off the waste heat is an important feature of the cryocooler. Liquid cooling is more effective than air or gas cooling because of high heat transfer coefficients and heat capacities achieved with liquids. Air cooling is attractive, on grounds of simplicity, to almost everyone concerned except those actually responsible for the cooling system. Frequently miniature machines are air cooled but are then located in a pod along with other heat emitting power equipment so the environmental temperature is increased. This complicates the cooling problem.

The *environmental temperature* over which the equipment is required to operate ranges from −40 °C to as much as 60 °C. This in itself presents problems which are compounded in many cases by the further requirement that the refrigeration temperature is a precise value having a tolerance as slight as ± 0.1 °C.

System cooling and heat rejection present special problems in space applications. All the heat must ultimately be rejected by radiation. The well-known equation:

$$Q = EAST^4 \tag{1.3}$$

where Q is the radiation heat flux, E is the surface emissivity, A is the area of the emitter, S is the Stefan–Boltzmann constant equal to 5.67×10^{-8} W m^{-2} K^{-4}, and, T is the temperature of the emitter, clearly dictates a high temperature. However the coefficient of performance relation:

$$\text{COP} = \frac{T_E}{T_C - T_E}$$

dictates the lowest possible maximum cycle temperature, T_C (corresponding to T in the above radiation equation).

Continuing with the special requirements for aerospace cryocoolers, important characteristics are the ability to withstand the high acceleration and vibration spectrum of a rocket launch, and the ability to operate in any orientation and in a zero or low gravity regime. Reliability and long life assume an importance not found elsewhere.

1.3 Classification

Low-capacity cryogenic refrigerators systems may be broadly classified into two principal groups as shown in Fig. 1.1:

(1) mechanical cryocoolers
 (i) open-cycle systems,
 (ii) closed-cycle systems;

(2) Non-mechanical cryocoolers.

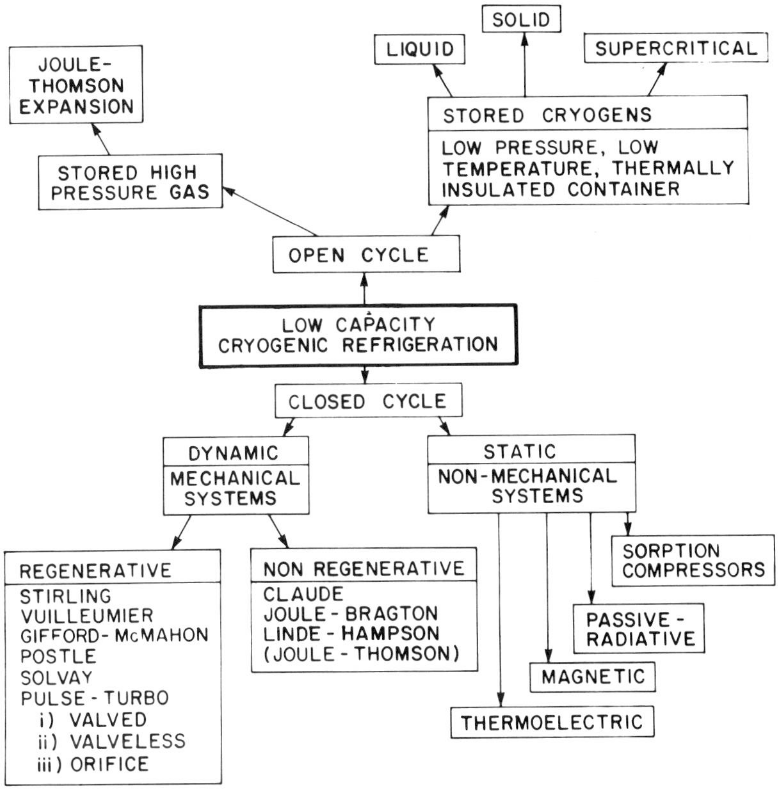

Fig. 1.1 Classification of low-capacity cryogenic refrigerators.

1.3.1 Mechanical cryocoolers

Open-cycle systems Open-cycle systems are those in which the working fluid is expendable. A finite quantity of the working fluid is provided and cryogenic cooling is obtained with very high reliability until the cryogen is exhausted.

There are two basic open-cycle systems:

(1) stored gas at high pressure and ambient temperature expanding in a Joule–Thomson process to obtain cooling;

(2) stored cryogens at low temperature and ambient pressure in highly insulated storage flasks or containers (dewars). The cryogens may be stored as:

(i) liquid
(ii) solid
(iii) liquid–solid mixture (slush)
(iv) supercritical fluid.

Closed-cycle systems Closed-cycle systems may be broadly classified as:

(1) mechanical refrigeration systems, that is, dynamic systems;

(2) non-mechanical systems, that is, static systems.

The mechanical refrigeration systems may be further subdivided into two groups depending whether or not they include a regenerative heat exchanger as follows:

(1) regenerative;

(2) non-regenerative.

Regenerative mechanical systems are by far the largest group of low-capacity cryogenic refrigerators. They may be further subdivided according to their ideal thermodynamic cycle of operation as follows:

(1) Stirling;

(2) Vuilleumier;

(3) Gifford–McMahon;

(4) Solvay;

(5) Postle;

(6) Pulse-tube.

These different cycles are found in a surprisingly diverse range and variety of mechanical configurations and refrigerating capacities from fractions of a watt (at 80 K) to kilowatts of refrigeration (at 80 K). Some of these variations are explored in subsequent chapters.

A critical feature of any arrangement is whether or not the cold expander part of the engine is separated from the compressor part by a long, thin, transfer line. Machines where the compressor and expander elements are found in the same assembly are said to be the 'integral' type, for example, integral-Stirling. Machines where the two elements are separated are said to be 'split', for example, split-Stirling. This feature has an important effect on the suitability of one machine or another for particular applications. In general, the integral machines are more efficient but split machines allow the possibility of a small lightweight cold region and, moreover, allow the compressor with its attendant vibration and electromagnetic emissions to be located some distance from the device being cooled.

1.3.2 Non-mechanical systems

Thermoelectric refrigerators are small, lightweight static systems with no moving parts. Their potential for long-term operation without maintenance is very high but at present thermoelectric refrigerators are unable to achieve cryogenic temperatures even when multistaged. This awaits the development of materials possessing the appropriate physical properties. Nevertheless their inclusion here among the ranks of cryogenic refrigerators is justified because of the great advantage of no moving parts and the reasonable prospect of improved materials.

Magnetic refrigerators are at a relatively advanced stage of development while not yet in production or widespread commercial usage. They come in various forms, some of which do include moving parts, others where there are no moving parts. The decision to include them under 'non-mechanical systems' is therefore arbitrary and open to revision.

Sorption compressors are used to compress gases and have no moving parts apart from inlet and outlet valves to regulate the gas flow. They utilize the process of adsorption or absorption of gases by selected media with subsequent release of the gas at elevated temperature. Such compressors are at an advanced stage of development although none are yet in commercial production.

Radiant coolers are used for some space applications. They use the phenomenon that deep space has a very low virtual temperature so the radiation energy received from space is negligible. Thus a radiation panel equipped with a high-emissivity surface can radiate energy to space and in so doing achieve cryogenic temperature levels. Of course, it is necessary for the radiant surface to be carefully shielded from all radiation and conduction thermal inputs from adjacent spacecraft surfaces or other bodies such as the Earth or Moon. Radiant coolers have been used for spacecraft instrument cooling for many years. The refrigeration capacities that can be achieved are very limited and temperatures less than 80 K are rare.

1.4 Organization of the book

Chapter 1 introduces the broad field of low-capacity cryogenic refrigeration and establishes some of the nomenclature, the jargon of the technology, and attempts to classify the various systems available.

Chapter 2 discusses, briefly, different applications for low-capacity cryogenic refrigerators, some well-established needs and requirements, some that are presently developing, and others that are yet in prospect.

Chapter 3 covers open-cycle systems with expendable working fluids.

Chapter 4 embraces the closed-cycle mechanical systems which include a regenerative heat exchanger. Chapter 5 covers the closed-cycle mechanical systems which have only recuperative heat exchangers.

Chapter 6 includes the static systems, thermoelectric and magnetic refrigeration, radiant coolers, and sorption compressors.

Chapter 7, contributed by Dr Otto Ledford, reviews the history and present status of mechanical cryorefrigerators in spacecraft.

Chapter 8 reviews the available sources of information, books, periodicals, conference proceedings and the like.

The appendix contains the names and addresses of organizations, companies, government departments, and universities with an active interest in low-capacity cryogenic refrigeration. Three indexes are included, a subject index, an author index, and an institution index.

1.5 Closure

In this introductory chapter we have reviewed both the nomenclature peculiar to low-capacity cryogenic refrigeration and some of the important parameters and operating characteristics. A classification scheme for the many different types of refrigerators or cryocoolers was introduced and then incorporated into the organization structure of the book. First we look at some of the applications, followed by technology surveys of the most widely used methods to achieve low-capacity refrigeration, some alternative systems presently in development, the special field of space applications and, finally, a review of the principal literature of low-capacity cryogenic refrigeration, sources of information, and a limited directory of the principals active in the field.

References

Smith, J. L., Robinson, G. Y. and Iwasa, Y. (1984). *Survey of the State of the Art of Miniature Cryocoolers for Superconductive Devices*. Office of Naval Research Contract No. N0001483K0327. Cryogenic Engineering Lab., Department of Mechanical Engineering, Massachusetts Institute of Technology, Cambridge, Mass.

Strobridge, T. R. (1974). *Cryogenic Refrigerators—An Updated Survey*. National Bureau of Standards Technical Note 655, Superintendant of Documents,

US Government Printing Office, 12 pp. (Charts reprinted in Walker, G., (1983) *Cryocoolers*, 2 vols., Plenum, New York.)

Voth, R. O. (1974). *Cryogenic Refrigerators for Shipboard Forward Looking Infrared Applications*. NBSIR 74–372, National Bureau of Standards, Boulder, Colo.

2

APPLICATIONS FOR LOW-CAPACITY CRYOGENIC REFRIGERATION

2.1 Introduction

There are many and diverse applications for low-capacity cryogenic refrigeration in military, civil, medical and scientific fields both existing and in prospect. The temperatures of interest range from less than 1 K to temperatures well above the cryogenic limit (110–120 K). Required refrigeration capacities range from fractions of a watt to well over a kilowatt. In some cases precise temperature control maintained to very close limits is very important. In others mechanical vibration or the electromagnetic emissions of the drive motor must be minimized. Some situations require a very high order of reliability with low or zero maintenance requirement. Yet others are concerned to minimize the power input to the system, have special concerns about a low mass of the expander unit or need to carry out the cooling process at environment temperatures both above and below the customary value. Some applications combine multiple priorities providing a challenge to the cryocooler designer and manufacturer that is frequently hard to meet.

2.2 Infrared detectors

One of the earliest needs for cryogenic cooling and, in terms of numbers of individual units made and used, by far the largest present application, is to cool the sensors used in infrared detectors. They can be used at night as well as in daylight to observe warm objects and are unaffected by weather, smoke, and so on. They are indispensable for certain tactical and safety-related functions.

The main application of infrared systems at present is military in night vision equipment and missile guidance systems. However, there are increasing civil applications for infrared sensing and thermal imaging. These range from night vision equipment for security, police and border patrol work, fire detection alarm systems, energy conservation thermographic analysis, medical thermographic examination for early cancer detection, astronomical observation, earth resource mapping, and process control instrumentation for observing and detecting, say, the liquid level

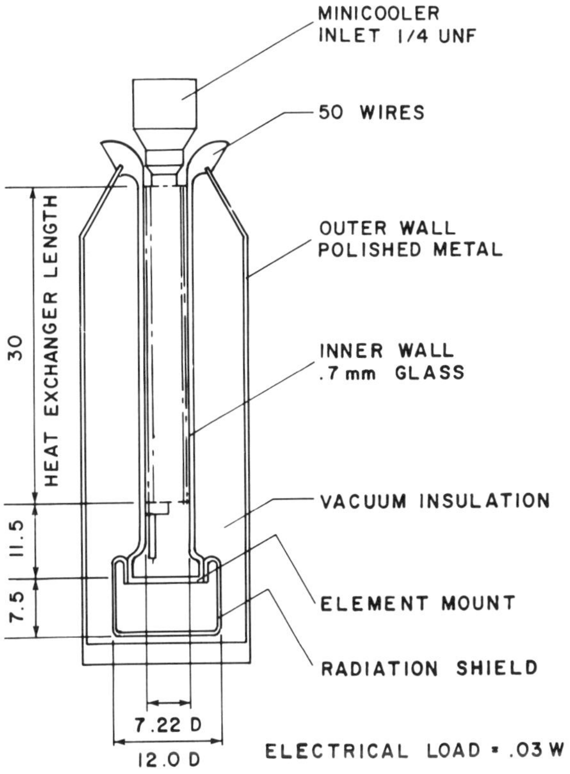

Fig. 2.1 Encapsulation detail for a multielement infrared detector and Joule–Thomson minicooler (courtesy Hymatic Engineering Co. Ltd.).

in storage tanks. These applications are anticipated to increase in great numbers as costs are reduced and reliability increased.

For military applications alone miniature cryogenic refrigerators are already in series production at the rate of several thousand units a month for night vision and missile guidance ('heat-seeking') systems. The development of 'smart' munitions will increase this many times over. Future artillery shells will have small guidance control surfaces which flip out on leaving the gun barrel or launching tube to be used for steering by an elementary heat-seeking guidance system located in the nose. Similarly, guided projectiles and beam-guided weapons may also be equipped with infrared heat-seeking sensors.

For many infrared detector applications the multielement detectors on a silicon or epoxy–fibreglass mounting platform are encapsulated as shown in Fig. 2.1 within a glass or stainless steel vessel or dewar flask (after Sir James Dewar, a famous nineteenth-century Scottish scientist and early cryogenics pioneer).

The detectors are mounted inside the dewar at the closed end of a long thin tube extending coaxially into the dewar. This is the *cold well* into which the *cold finger* of the refrigerator fits. The tip of the cold finger is cooled to cryogenic temperature, usually around 80 K, and this cools the closed end of the cold well carrying the infrared detector thereby cooling the detector. The *interface* between the end of the cold finger and the bottom of the cold well is the source of considerable difficulty we shall discuss later.

Various *leads* of very fine copper wire are attached to the detector and then loosely coiled within the vacuum space to provide a long thermal conduction path before attachment to a multiple-pin *pass-through*. This permits the electrical connection of the detector elements to the electronics unit translating the detector signals into a television picture.

A *window* is provided at the end of the dewar to allow passage of the incoming radiation to the infrared detector.

Mirrors and lenses focus the incoming radiation on to the detectors. For small systems a few detectors are used and a small oscillating mirror *scans* the view ahead of the sensor projecting parts of the image sequentially on to the detector. Bigger systems with a large array of detector elements do not have scanning equipment. Instead they *stare* at the whole scene ahead and provide a continuous signal of the whole image.

Infrared detectors are essentially heat-sensing instruments and must be cooled to respond to the sources of interest. Typically, detectors operate in the temperature range 100 K to 4 K with the cooling requirements ranging from two or three watts at the upper temperature to fractions of a watt at the lower temperatures.

For military applications, sensing people and warm objects such as engines or power systems, the infrared detectors sense emissions in the range 1–25 μm. There are basically two types of device:

(1) quantum, including photoconducting and photovoltaic;

(2) thermal, including bolometric and pyroelectric—dipole and thermocouple.

The quantum devices are more sensitive whereas the thermal devices are more linear (Fieldhouse and Porter 1986).

With increasing demands on sensitivity, noise is a particular problem. Noise in infrared systems is due to:

(1) photon noise, the so-called dark current;

(2) detector noise due to thermal agitation;

(3) flicker or contact-surface noise;

(4) differential temperature;

(5) vibration, or so-called microphonics.

An important figure-of-merit parameter of infrared devices is the *detectivity*, D, the reciprocal of the noise equivalent power, NEP. This is the incident power resulting in a signal to noise ratio of unity. Contemporary 256- and 512-element arrays have NEP values of 3–4 $\times 10^{17}$ watt when cooled to temperatures in the range 25 to 80 K and provide resolution to 0.03 K (Fieldhouse and Porter 1986).

The detectivity D is often expressed theoretically in terms of the area A and band width, df, as represented by $D^* = D(A\,df)^{\frac{1}{2}}$. A reduction in temperature of the infrared system increases D^*.

When evaluating a refrigerating system for an infrared system it is important to consider how the cold-end temperature will affect the thermal sensitivity, the spectral transfer function, and the modulation transfer function of the system. This is related to the detectivity, responsivity, noise spectra, spectral response, and frequency response of the cooled array (Voth 1974).

Detectors of mercury–cadmium–telluride ($Hg_{(1-x)}\,Cd_xTe$) are widely used as photoconductive infrared sensors. The value x in the above description varies, with resulting variations in the spectral response and other characteristics. A typical value of $x = 0.2$ produces a spectral response shown in Fig. 2.2 with the detector cooled to 80 K (after Voth 1974). This is the customary temperature range of operation for many infrared detectors operating with liquid nitrogen bath cooling or, as in many cases, cooling by the liquid nitrogen or liquid air resulting from the Joule–Thomson expansion of compressed nitrogen or air.

The effect of temperature on the detector responsivity and noise characteristics are shown by Fig. 2.3. This shows the specific detectivity (D^*) as a function of the detector temperature. It can be seen that at temperatures greater than 80 K, D^* decreases rapidly, whereas between 60 and 80 K there is relatively little change. The spectral response to shorter wavelengths as the temperature increases defines the temperature region 60–80 K as a particularly suitable temperature regimen for mercury–cadmium–telluride detectors.

Many details of the various types of detectors and their particular technology have been given by Hudson (1969), Lloyd (1975), and by Wolfe and Zissis (1978). Contemporary developments are frequently

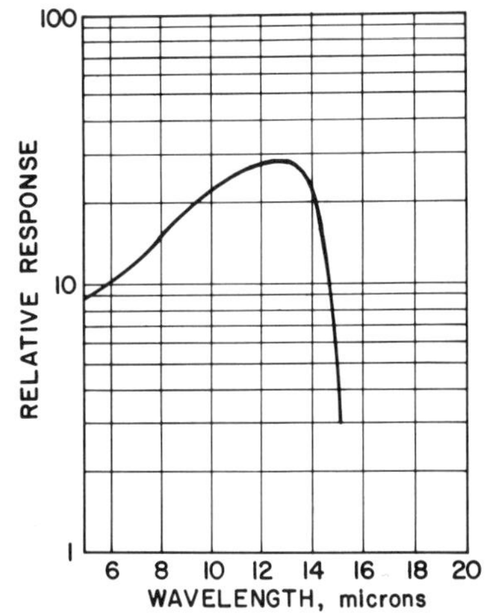

Fig. 2.2 Typical spectral response of a mercury–cadmium–telluride detector cooled to 80 K (arbitrary units) (after Voth 19741).

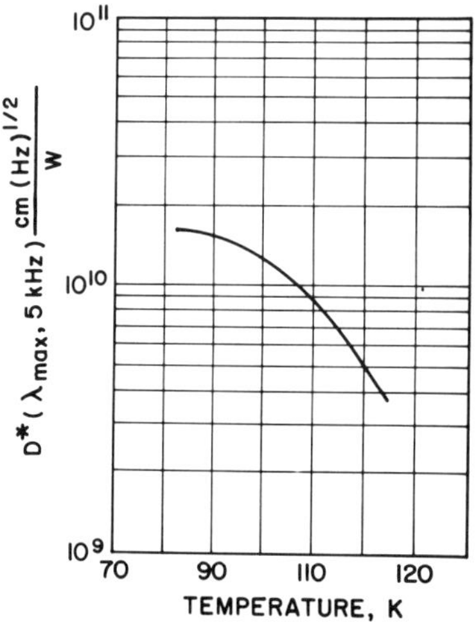

Fig. 2.3 Specific detectivity of a typical infrared detector as a function of detector temperature.

reported in the optics industry journals and conference proceedings, such as, the SPIE Conference Proceedings of the Society of Photo-Optical Instrumentation Engineers.

In general there is an inverse relationship between the maximum wavelength of the radiation and the temperature of operation of the infrared detector. A sensor on a weather satellite responding to relatively short-wave infrared radiation will operate satisfactorily in the region of 200 K. An astronomical satellite responding to very long-wavelength radiation must be cooled to very low temperatures near absolute zero. Jamieson (1984) indicated many space applications, both strategic and scientific, which require cooling in the range 15–80 K with an extreme of 2.8 K for the Infrared Astronomy Satellite.

The US National Aeronautics and Space Administration (NASA) has two main areas of observational scientific endeavour, astrophysics and earth science, with requirements for sensors and detectors of high energy resolution. With present sensor technology NASA's goals can only be achieved by cooling sensors and instruments with stable operating temperatures extending from ambient down to 0.1 K or less.

Sherman (1982, 1983) reviewed NASA's needs for cryogenic cooling systems. More recently Castles (1988) and Castles *et al.* (1993) have reviewed progress and redirection in the development of cryocoolers to meet NASA needs. Some of the projects and facilities have lifetimes of up to 10 or 15 years, so long life or the capability to operate with limited servicing while providing a high degree of reliability are important aspects of the NASA program.

2.3 Cryopumping

Various industrial, commercial, and scientific processes involve the requirement for very high-vacuum conditions to be established quickly, reliably, and routinely with little or no expensive equipment or particular expertise.

This need has provided an opportunity for mechanical cryocoolers capable of achieving temperatures of 20 K or below. These are used in conjunction with a mechanical 'roughing' pump which exhausts to a low pressure the chamber to be evacuated. The gases and vapours remaining in the chamber after this preliminary stage are precipitated on the cold finger and shields of the cryocooler. This is strategically located so as not to interfere with the process (such as metal sputtering or vapour deposition) for which the chamber is being used.

The cryopumping vacuum process is particularly attractive for industrial processes where oil contamination of surfaces cannot be tolerated. It has been widely adopted by the silicon-based semiconductor industry.

A typical design for a small cryopump based on a two-stage cryocooler (invariably of the Gifford–McMahon type) is shown in Fig. 2.4. The

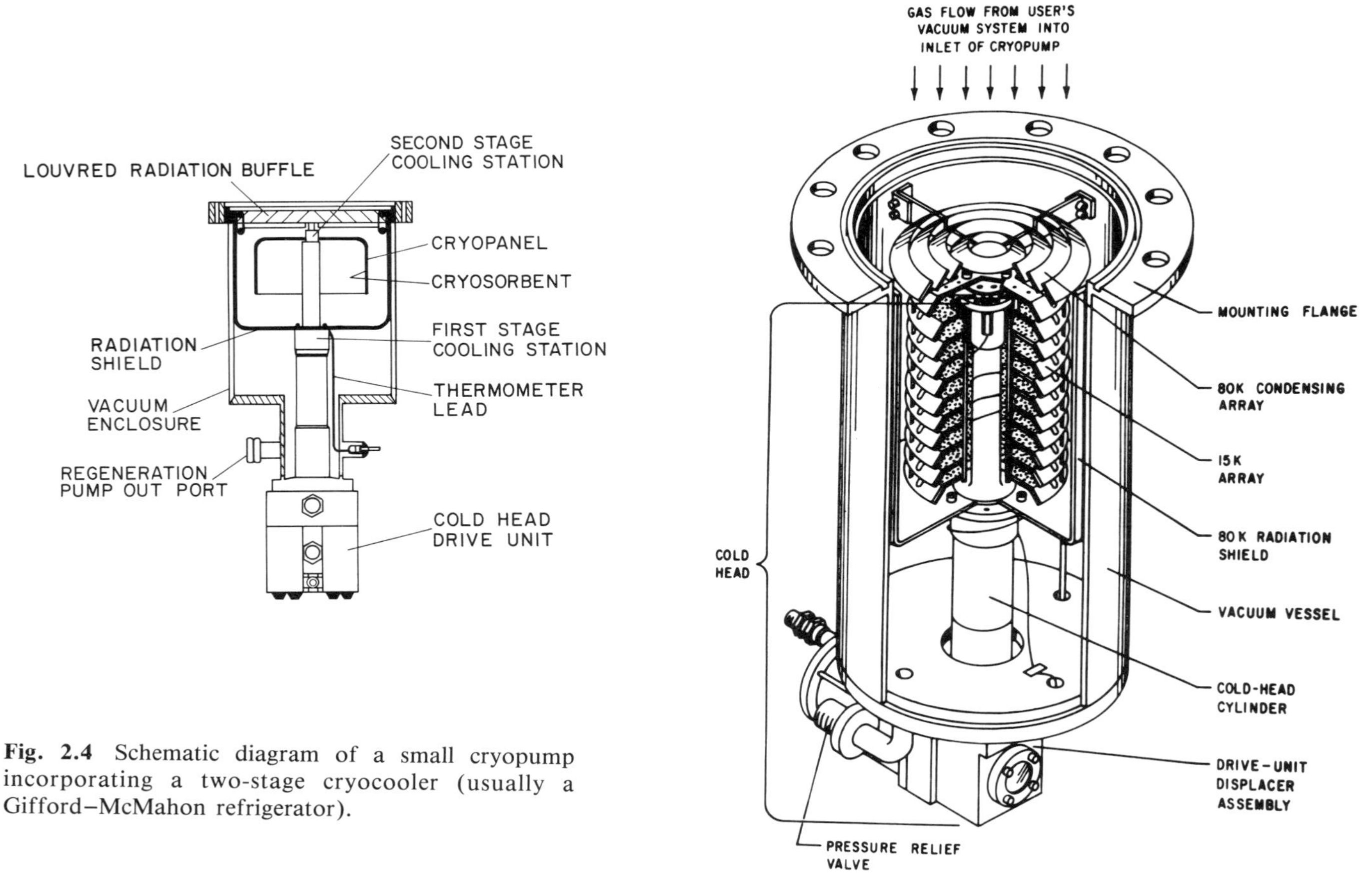

Fig. 2.4 Schematic diagram of a small cryopump incorporating a two-stage cryocooler (usually a Gifford–McMahon refrigerator).

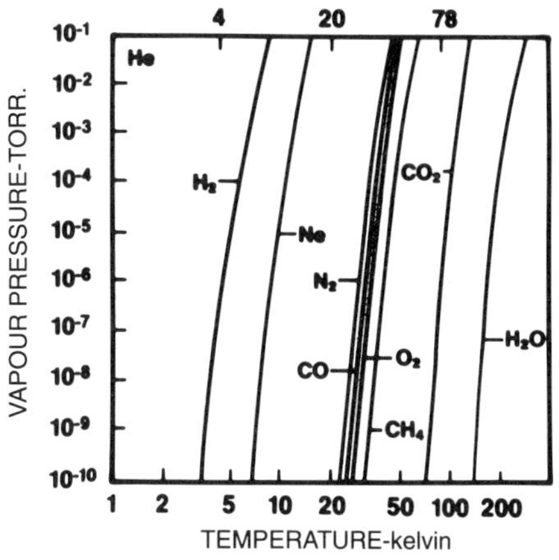

Fig. 2.5 Vapour pressure of some of the constituent gases found in atmospheric air as a function of temperature (after APD Cryogenics Inc.).

cryocooler produces refrigeration at two temperature levels, say 20 K and 77 K At the upper level a louvred or chevron baffle or cold trap is coupled by the radiation shield to the upper-temperature refrigeration stage but is remote from it so the baffle temperature is well above the 77 K temperature. However, the baffle is sufficiently low in temperature for water vapour and carbon dioxide to precipitate upon it. A cryopanel coupled to the lower-temperature refrigeration stage, and completely enclosed within the radiation shield, is cooled to a sufficiently low temperature for nitrogen, oxygen, and argon to be precipitated. Furthermore, an adsorbent† (cryosorber) mounted on the back (inside) of the cryopanel adsorbs hydrogen and helium. Sorbent materials include activated charcoal and zeolite also known as 'getters'. They remove gases by a combination of absorption and adsorption processes. Getters become increasingly effective at low temperatures. Single. stage sorption processes (cooled to 80 K) achieve pressures down to 100 mPa (about 10^{-3} torr)†† and with further cooling (to 20 K) achieve pressures as low as 100 μPa (about 10^{-6} torr) (Barron 1985).

† When gas is brought into contact with a solid some of the gas is usually taken up by the solid. When the gas molecules penetrate into the solid the process is called *absorption*. If the gas molecules stick to the surface of the solid in one or more layers the process is called *adsorption* (see Barron 1985, p. 227).

†† The torr, a unit of pressure widely used in vacuum engineering, is defined as 1/760 of the standard atmospheric pressure, or for all practical purposes, 1 mm Hg absolute. In SI units 1 torr = 1/133 Pa or 3 kPa = 25 torr.

Vapour pressures of some of the constituent gases found in atmospheric air are shown as a function of temperature in Fig. 2.5.

The sorbent material eventually becomes saturated and the gas adsorped or adsorbed must be desorbed. This is normally accomplished by allowing the sorbent material to attain room temperature or above (to as much as 250 °C to release adsorbed water vapour).

2.4 Medical instrumentation

Instruments and devices involving the use of cryogenic systems are being used increasingly in hospitals and medical clinics. Sarwinski (1988) has reviewed the cryogenic requirements for the principal equipment. Kelley (1991) includes reviews of various applications of cryogenics in the medical and biological field.

2.4.1 Magnetic resonance imaging

Magnetic resonance imaging (MRI) systems are the largest, and most common, cryogenic instruments used for medical purposes. Sarwinski estimated (in 1987) there were about 600 in operation world-wide in both fixed and mobile installations. Their use is increasing quite rapidly so it would be reasonable to assume that the number has increased, to something well over 1000 (in 1993).

MRI machines are used to obtain a two-dimensional picture displayed on a cathode ray screen of the hydrogen distribution in a biological sample. The picture is interpreted by a specialist as showing the internals of a body, somewhat akin to an X-ray picture but without the need for potentially harmful X-rays.

An MRI installation consists of a superconducting magnet contained in a toroidal vessel having an inner bore of 1 metre, a length of 2 metres and a mass of up to 10 tonnes. The magnet is immersed in a liquid helium bath of 300 to 1500 litre capacity. This is provided to cool the magnet to superconducting temperatures necessary to achieve the intense magnetic fields required.

The helium reservoir is contained in a vacuum space with a radiation shield cooled to 30 K by the cold helium vapour generated from the helium bath. The cold radiation shield is itself enclosed by a liquid nitrogen bath or another shield cooled to liquid nitrogen temperatures. The nitrogen reservoir may contain 30 to 300 litres of liquid nitrogen. Superinsulation is used to reduce the heat lost from the outside to the cooled shield.

Typically liquid helium is lost by evaporation at the rate of 0.3 to 1 litre per hour and needs to be replenished every month or so. Liquid nitrogen is lost by evaporation at the rate of 0.5 to 1 litre per hour. Since the volume of nitrogen is usually much less than helium the nitrogen must be replenished more frequently, every day or so. This may be accomplished automatically by transfer from an external storage reservoir.

Replenishment of the liquid helium is more difficult and must be accomplished by a skilled technician. Even then a considerable loss (by evaporation) of helium occurs. When attempted by inexperienced personnel, the loss can be substantial. Presently, in North America, helium costs about $5 per litre (in substantial quantities) so that the loss of 500 litres represents a loss of $2500. In other parts of the world liquid helium and nitrogen are not so readily available and are more expensive. The services of skilled operators may also be difficult to engage.

These considerations have prompted the use of cryocoolers to reduce the consumption of helium and nitrogen. Some systems retain the large helium and nitrogen reservoirs and use the Collins cryostat, a Claude cycle helium liquefier, to recondense the helium boil-off vapour with an additional circuit of helium at 70 K circulating in the liquid nitrogen to eliminate the evaporation of nitrogen. Retention of substantial volumes of the cryogenic liquids increases the temperature stability of the magnet. It also allows the MRI instrument to remain in operation during scheduled maintenance or emergency shut-down of the cryocooler. Usually repair by replacement of the cryocooler prevails.

The use of an on-site liquefier allows operation in any part of the world for it relieves the problem of securing and transferring the liquid cryogens. However, its use does introduce the problem of maintenance and reliability of the liquefier. Present helium liquefiers are relatively complicated machines that need to be monitored in operation and maintained at regular intervals.

Furthermore, an on-site liquefier represents a considerable investment (about $120,000 compared with approximately $1 M for the MRI unit. The smallest machines presently available, at the minimum capacity rating, produce far more cooling than is necessary, about 5 watts of cooling at 4 K (corresponding to 15 litres of liquid helium per hour) and 60 watts at 80 K. Smaller Stirling cycle machines of more appropriate capacity (8 to 12 litres per day) are in development (Ellison *et al.* 1988) and strenuous efforts are being made for these liquefiers to operate reliably for long periods without maintenance. Their relatively small size and volume will permit their direct mounting to the helium dewar.

Gifford–McMahon cryocoolers are also used to minimize the need to replenish the cryogens. Longsworth (1984, 1986) has described various systems ranging from (a) a simple single-stage machine cooling to liquid nitrogen temperatures to maintain the nitrogen reservoir, (b) a more complicated two-stage system cooling both the 80 K and 30 K shields (perhaps to a lower temperature, 15 to 20 K), and (c) a two-stage liquefier with a Joule–Thomson expansion stage to liquefy the helium boil-off vapour at 4 K. Multiple cryocoolers may be installed to improve reliability or to increase the capacity where installed.

2.4.2 Positron emission tomography

Positron emission tomography (PET) is another advanced medical technique requiring cryogenic refrigeration (Sarwinski 1988). In this technique multiple detectors determine, with great precision, the location of positron-emitting radio nuclides. Computers resolve the detector signals into images displayed on video screens.

The radio pharmaceutical source materials have short half-lives (minutes to hours). They are produced as close as possible to the point of use in small, mobile cyclotrons. These utilize superconducting magnets and so require cooling with liquid helium. All the cryogenic considerations about magnetic resonance imaging apply also to positron emission tomography. An additional requirement is that any supplementary cooling provided by an external cryocooler must penetrate the heavy iron and lead shielding of the cyclotron unit.

2.4.3 X-ray scanners

Certain crystals, when cooled to low temperatures, are used as sensitive, low-noise X-ray detectors. They require cooling in the range 80–130 K which may be accomplished using liquid nitrogen rather than helium. A capacity of about 5 watts at 80 K is required and is invariably provided by a liquid nitrogen bath with consequent loss of coolant by evaporation.

Mechanical cryocoolers cannot be used to reliquefy the vapour boil-off because of their vibration. The alternative Joule–Thomson expansion nozzles potentially useable are not available commercially although recent developments in closed-cycle Linde–Hampson cryocoolers (Hedegard *et al.*, 1989) suggest these units may soon be available.

Fortunately replenishment of the liquid nitrogen bath requires much less skill and experience to accomplish than liquid helium and is much less costly.

2.4.4 SQUID instruments

*S*uperconducting *QU*antum *I*nterference *D*evices (SQUIDs) are highly sensitive magnetometers. They have been used in research work for the past 20 years or so for the investigation of physiological magnetic fields. They are now passing into clinical diagnostic use as a non-invasive procedure for studying the major organs of the body.

The basic SQUID instrument has one or more signal channels each associated with a superconducting pick-up coil contained in a non-metallic open dewar of 10 to 15 litres of liquid helium. The dewars generally have radiation shields cooled by helium vapour to avoid the use of supplementary liquid nitrogen cooling.

The liquid helium must be replenished about once a week, usually from a larger storage dewar at the site. The refilling process requires the services of a skilled technician. It is a process that Sarwinski (1988) and, presumably, those engaged in the transfer process, feel could benefit considerably from the development of superior apparatus and technique.

Helium boil-off rates are of the order of 1.5 litres per day from the SQUID instrument and 1 litre a day from the storage dewar plus the losses incurred in the refill process. This limits to about a month the time a 100-litre liquid helium storage dewar can service a 15-litre instrument.

As these instruments pass into more general use the logistical problems of supply of the storage dewars and the availability of technicians with the necessary skills to accomplish the weekly transfer will become acute.

The situation could be greatly alleviated by the use of a low-capacity helium vapour recondenser. At present such units are not available commercially but several systems are in development. Direct coupling of a mechanical cooler to reliquefy the helium or reduce the boil-off is difficult and complicated because of the acoustic and magnetic noise generated by the electric drive motor and the reciprocating parts.

Recently Buchanan *et al.* (1988) discussed the use of a commercial two-stage Gifford–McMahon refrigerator with a 4 K Joule–Thomson (JT) stage added at the bottom to cool a SQUID magnetometer system. It operated successfully at equivalent noise levels of 2.5×10^{14} $T/\sqrt{\text{Hz}}$ over a bandwidth of 1–500 Hz except at a few discrete frequencies. These noise spikes were then filtered with a computer-based adaptive filter. The elimination of liquid cryogens in this system permits its operation in any orientation.

Sarwinski (1988) favours the use of a cascade JT system mounted on the instrument. This would produce almost no vibration at the cold end and has no moving parts to interact magnetically with the SQUID pick-up coil. Unfortunately, cascade JT systems are not yet available commercially although, again, efforts to develop low-capacity are in progress with promising results. The cascade JT system uses helium to achieve the lowest (4 K) temperatures but the compressed helium must be precooled before expansion. This can be done using another circuit containing hydrogen which itself must be precooled by a third system containing nitrogen or argon. With three fluids all operating in closed cycle Linde–Hampson systems and including separate (perhaps ganged) compressors a considerable investment is involved. Moreover, the services of a site engineer to monitor and maintain the system may still be required although developments in novel compressor systems may render this unnecessary.

One particular difficulty of a cascade JT system is that one of the fluids would likely be hydrogen (used to precool the compressed helium and itself precooled by nitrogen). Compressed hydrogen is very difficult to contain over the long term and its escape, even in minute amounts, poses a fire and explosion hazard.

2.5 Cryobiology

Cryobiology is the name given to the area of biological technology involved with the application of very-low-temperature procedures and techniques, principally cryosurgery and organ preservation. Cryopreservation is concerned with the arrest and recovery of living processes of cells, tissues, and organs. Cryosurgery is concerned with the local cryodestruction of diseased cells while preserving the healthy surrounding tissue. The technological requirements of the cryobiologist are very demanding: the ability to cool tissues very fast (10^6 degrees per sec, the process of vitrification); to thaw frozen tissue without damaging the delicate cells, to freeze-dry tissue using molecular distillation (vacuum) drying, to supercool cell structures below 0 °C without freezing, and to successfully store the preserved tissues and organs for any required length of time.

There is a Society for Cryobiology and in concert with the Cryogenic Society of America they sponsored CRYO–90, an unusual interdisciplinary conference drawing from the life sciences as well as the physical sciences and engineering areas of the cryogenic community. The proceedings of this conference (Kelley 1991) are a good summary of the state of development of this interesting field.

2.6 Cold electronics

Low-temperature operation of electronic systems is being increasingly applied or contemplated for a wide range of devices and systems. These range from miniature single-transistor amplifiers to large power systems employing many very large scale integrated (VLSI) circuits for ultra-fast 'supercomputers' (Kirschman 1985).

Operating electronic systems at low temperature provides the means to extract better performance from existing technology; things work faster and better with less noise. It avoids the expense and delay involved in waiting for advances in design and fabrication to achieve the same result at room temperature. It also provides an opportunity to take advantage of effects which are available only by operation at low temperature thereby stimulating the development of new, specialist, devices based on these effects.

Electronic systems can be broadly subdivided into two classes: semiconductive electronics and superconductive electronics. Operation at cryogenic temperatures is presently essential for superconducting systems (although some believe room-temperature devices may be possible in the future) whereas it is an option for semiconductor devices.

2.6.1 *Semiconductor electronics*

Kirschman (1985) asserts that the use of semiconductor materials at low temperatures reflects the situation at ambient temperatures. The principal material is silicon (Si) with increasing usage of gallium arsenide (GaAs) and a few germanium (Ge) devices for specialized purposes.

Nearly all the cooled semiconductor devices are of the field-effect transistor (FET) family, including Si and Ge junction FETs (JFETS) and Si metal oxide semiconductor FETs (MOSFETs).

Temperature requirements range from liquid helium (4 K) to over 100 K and up to room temperature. There is a focus of interest at 80 K because of the ease and ready availability at low cost of liquid nitrogen. For many devices the gain in performance at lower temperatures is not worth the additional cost and complication involved.

The switching speed of semiconductor devices can be increased by lowering their temperature. As individual device speed increases, the interconnections between devices on an integrated circuit (IC) chip also become important in determining the speed of complex digital circuits. As ICs become denser and at the same time larger, the interconnecting lines are longer with smaller cross-section. Reducing the temperature improves performance because it reduces the resistance of the interconnecting materials.

To achieve high speed the IC circuit elements must be packed close together, but this increases the power density and complicates the heat removal problem, the limiting factor in circuit performance at room temperature. Benefits in this regard can arise from improved cooling (immersion in a liquid nitrogen bath).

Moreover, operation at low temperature reduces the power dissipation necessary with dense circuits. Reduction of the operating voltage reduces the power level and this can be accomplished by reducing temperatures so as to reduce the thermal voltages. Leakage in p–n junctions and insulators is reduced at low temperatures so less power is required to store and maintain data.

Noise is very much reduced at low temperatures. The effect is progressive but, in many cases, not very significant at temperatures below 80 K.

Nearly all the degradation mechanics in electronic systems are related to temperature. The well-known sharp decline in reliability consequent on increased temperature implies an order of magnitude (or more) improvement in operating life at lower temperature although this has yet to be demonstrated. Some miniature devices presently in development may have such a high failure rate at room-temperature operation, or even storage, as to be useable only at reduced temperature.

One problem of intermittent operation at room temperature is the reduced reliability arising from the mechanical stresses and strains resulting from thermal cycling, repeated cooling and warming of the electronic chip. Reliability would likely increase with cooling to cryogenic temperatures.

The cooling capacity required with electronic systems varies, obviously, with the size, capacity, and power level. Small single-transistor amplifiers need only milliwatts of cooling at 80 K, even less at lower temperatures. Cryogenically cooled supercomputers have the highest capacity requirements. They incorporate many very large scale integrated circuits. The baseboard of these units, a thin fibreglass–epoxy plate about 1 metre square and packed with electronic chips, operates fully immersed in a bath of liquid nitrogen, with a total power dissipation from the board of 700 watts or so. Computers with multiple boards are being made. The largest known to the authors had 34 boards, with a total cooling capacity requirement of nearly 25 kW at 80 K, a sizeable cryogenic facility and well beyond the arbitrary limit of 1 kW at 80 K adopted as the limit of our interest here.

2.6.2 Superconducting electronics

Superconductivity is the phenomenon wherein metals and alloys lose their electrical resistance and so become superconducting. The transition to a condition of no electrical resistance occurs only at cryogenic temperatures and is a sudden transition taking place abruptly over a small temperature change (less than 1 K). The temperature at which it occurs depends on the local magnetic field intensity, geometry, and other factors.

The maximum transition temperature, or critical temperature, T_C, occurs in the absence of a magnetic field and varies with different materials over a considerable range. Critical temperatures greater than 10 K are arbitrarily defined as high. The usual operating temperature (for switching and other applications involving the presence of a magnetic field) is about half the critical temperature.

Until recently niobium alloys exhibiting maximum superconducting transition temperatures approaching 20 K were the highest-temperature superconductors known. However, in 1986 certain copper oxides were found to exhibit superconducting behaviour at substantially higher temperatures, 90 K and above. This has led to a veritable frenzy of research effort and is expected to result in further substantial elevation of the superconducting temperature limit. There are those who talk, even, of room-temperature superconductors although others regard this as unlikely, at least for the next few years. There is more or less general agreement that useful superconducting devices capable of operating at liquid nitrogen temperatures will be available relatively soon (that is, in a few years).

One of the principal areas of interest in superconducting electronics is the family known as Josephson devices exhibiting the behaviour predicted by B. Josephson of Cambridge University in the early 1960s. This entails a pair of superconducting metals separated by an insulating layer thin enough to permit two forms of electronic tunnelling. In the non-superconducting mode single-electron tunnelling requires a small voltage drop, typically a few millivolts, across the insulator. The alternative, superconducting paired-electron tunneling mode occurs with no voltage drop at all.

The conversion from one mode to the other can be influenced by changes in the local magnetic field. The two types of tunnelling and the ability to switch from one to the other by the application or withdrawal of a small magnetic field is the basis for the operation of the family of Josephson devices used for switching or incredibly sensitive magnetometers.

Josephson devices and the other superconducting electronic systems, superconducting quantum interference devices (SQUIDS) are finding application in diverse areas biomedical research and clinical diagnostics, geophysical measurements (earth resource assessment and discovery), metrology, electromagnetic radiation detection, digital signal processing and computing, magnetic anomaly (submarine) detection, voltage standards, space applications, millimetre wave and infrared detection, and other laboratory measurements of various kinds. All require some part of the device or system to be cooled to cryogenic temperatures. In nearly every case liquid helium baths are used to achieve the superconducting temperature level. The comments made above under SQUIDS in the medical instrumentation sections with regard to liquid helium supply and replenishment apply here also with equal force.

Nisenoff (1981) pointed out that an increase in operating temperature from liquid helium (4 K) to 10 K, a factor of 2.5, would reduce the cost, weight, size, power input, and complexity of the refrigerator by an order of magnitude. Given a ratio of the operating to maximum superconducting transition temperature of 0.6, an operating temperature of 10 K requires critical temperatures of 16 K, well below the limits of niobium alloy superconductors. 10 K can be achieved relatively easily by multistage expansion regenerative cryocoolers of the Stirling, Vuilleumier, and Gifford–McMahon variety.

Of course, the recent developments of very high-temperature superconducting materials offers the possibility for operation at 80 K, liquid nitrogen temperature, and easily achieved by a single-stage expansion engine of the above family.

In situations where 4 K or 10 K SQUIDs or other superconducting devices have to be used for one reason or another it would still be advantageous to use 80 K superconducting wire in the fabrication of pick-up coils coupled to 4 K SQUIDS. This would be particularly attractive in some instruments where the pick-up coils have to be located as close as possible to the warm outer surface of the dewar.

2.6.3 High-temperature superconducting magnets

A principal application of the phenomena of superconductivity is in very high-intensity electromagnetic superconducting magnets. Presently the coils of these magnets are wound from wire made of niobium alloys.

The development of practical high-temperature superconducting magnet wire would substantially facilitate and extend the use of superconducting magnet systems. The general consensus of opinion in the development of such wire is that it may be 5–10 years away. The magnet wire would need to be ductile and flexible for forming and winding, able to be joined to itself and other metals or materials, have a high critical field, and high critical current with joints of low resistance to realize near-persistent operation. A major problem with the high T_C material is the low critical current densities that are presently achieved.

The availability of 80 K superconducting magnet wire would simplify the construction of many devices using superconducting magnets. For example, in the magnetic resonance imaging systems discussed above the intermediate vapour-cooled 30 K shield could be eliminated. An open bath dewar of 1500 litres of liquid nitrogen would still experience some heat leak from the outside and would need to be refilled periodically but with liquid nitrogen rather than liquid helium at a fraction of the cost and complexity. On-site reliquefiers of the nitrogen vapour are commercially available. They cost much less than helium liquefiers, consume much less power and are relatively simple, reliable machines.

2.7 Scientific research

There is a surprisingly wide and diverse requirement for low-capacity cryogenic refrigeration in scientific research. Description of the various applications may be found in the proceedings of the specialist cryogenic conferences, such as the Cryogenic Engineering Conference or the International Cryogenic Engineering Conference, and the specialist journals, such as *Cryogenics*. Requirements range all the way from fractions of a watt at the lowest temperatures approaching absolute zero to very large capacities involving substantial cryogenic installations. Special needs include very precise temperature control, the elimination of mechanical vibration and electromagnetic interference, an absolute reliability for a prescribed time, and so on. Many of these special requirements can be accommodated in open storage dewars with a liquid cryogen bath.

Much of this type of work involves the use of cryogens only incidentally as the most convenient way to achieve the low temperatures desired. The need to periodically transfer helium or refill liquid nitrogen bottles is simply accepted as a necessary evil, akin to the provision of basic services such as cooling water, electricity, compressed air, and so on. The availability of small, low-cost, reliable cryocoolers capable of operation without special expertise or training and for long periods with no maintenance would be a boon to many scientific projects.

2.8 Industrial and commercial applications

Similar comments to those under scientific research apply here. There is a surprisingly diverse range of requirements for low-capacity cryogenic refrigeration across a broad industrial and commercial front. The principal use of mechanical cryocoolers in industry is, without doubt, cryopumping, the use of two-stage expansion cryocoolers to achieve high vacuum conditions.

There is also considerable activity in the 'cryogenic soaking' of cutting tools in nitrogen-cooled chambers to improve life and wear characteristics. Close-fitting metal parts are sometimes 'shrunk' by immersion in liquid nitrogen to permit assembly to take place. Bull semen and other biological samples are flown about the world while preserved at liquid nitrogen temperatures. Liquid nitrogen is widely used in metallurgical, chemical, and food processing, and so on. The list of different uses and applications in the cryogenics industry and commerce is seemingly endless.

2.9 Closure

In this chapter we have attempted to survey, albeit superficially, some of the applications, special needs, and requirements for low-capacity cryogenic refrigeration. There is at present no well-established yardstick for measurement and assessment of the cryogenic industry. The *Cryogenics Handbook* by Law (1981) is a valuable measure of the range of companies and countries involved. However, it is increasingly dated and there is an urgent need for an extended, up-to-date replacement. Moreover, the *Handbook* made no attempt to measure the size of the cryogenic market or the range of customers and their requirements.

Two newsletters, *Cryogenic Information Report and Cold Facts*, the newsletter of the Cryogenic Society of America, attempt to inform readers of contemporary developments in the field. The scholarly journal '*Cryogenics*' presents papers of archival value and some information about contemporary developments.

Another measure of cryogenic activity may be the numbers of participants attending the specialist cryogenic conferences and short courses. Of these the most important are the biannual Cryogenic Engineering Conference in the United States; the biannual International Cryogenic Engineering Conference at various centres in the world, except the United States; the scientific conferences devoted to low-temperature work, LT 20, and so on, sponsored by various physics and chemistry societies; the biannual International Cryocooler Conference; the Applied Superconductivity Conference; and the specialist Commission meetings of the International Institute of Refrigeration, an organization based in Paris with meetings taking place in many parts of the world. Short coursers on cryogenic engineering and on cryocoolers have been held annually for many years at the University of California, Los Angeles. More recently short courses on cryogenic engineering have been organized by the American Society of Mechanical Engineers in conjunction with their annual conference.

References

Barron, R. F. (1985). *Cryogenic Systems*. 2nd edn, Monographs on Cryogenics 3, Oxford University Press.

Buchanan, D. S., Paulson, D., and Williamson, S. J. (1988). Instrumentation for Clinical Applications of Neuromagnetism. *Advances in Cryogenic Engineering* (ed. R. Fast), **33**, Plenum, New York.

Castles, S. H. (1988). Current Developments in NASA Cryogenic Cooler Technology. *Advances in Cryogenic Engineering* (ed. R. Fast), **33**, Plenum, New York.

Castles, S., Cygnarowicz, T., Boyle, R., Sparr, L., Cory, R., and Connors, F. (1993). NASA/GFSC Cryocooler Development Program. *Proceedings of the Seventh International Cryocooler Conference* (ed. T. Ludwigsen), Kirtland AFB, Albuquerque, N.M.

Ellison, W., Walker, C., and Zylstra, S. (1988). Cryocoolers for the New High-Temperature Superconductors. *Journal of Superconductivity*, **1** (2), Plenum, New York.

Fieldhouse, I. B. and Porter, R. W. (1986). *Cryogenic Cooling of Infrared Electronics: a State of the Art Review*. GACIAC SOAR 86–02, Guidance and Control, Information Analysis Center, IIT Research Institute, Chicago (AD-A192 318).

Hedegard, K., Walker, G., and Zylstra, S. (1989). A Non-Clogging, Temperature-Sensitive, Closed Cycle, Linde–Hampson Cryocooler. *Proceedings of the Fifth International Cryocooler Conference* (ed. P. Lindquist), Naval Postgraduate School, Monterey, Ca., Aug. 1988.

Hudson, R. D. (1969). *Infrared Systems Engineering*. Wiley, New York.

Jamieson, J. A. (1984). Infrared Technology: Advances 1975–1984. Challenges 1985–1994. *SPIE Seminar Proceedings*, vol. 510, Infrared Technology X, Society of Photo-Optical Instrumentation Engineers, pp. 56–67.

Kelley, J. P. (ed.) (1991). *Applications of Cryogenic Technology*. Plenum, New York.

Kirschman, R. (1985). Cold Electronics: An Overview. *Cryogenics*, **25** (3), 115–22, Butterworth, Guildford, UK.

Law, B. (1981). *Cryogenics Handbook*, IPC Science and Technology Press, Guildford, UK.

Longsworth, R. C. (1984). Interfacing Small Closed Cycle Refrigerators to Liquid Helium Cryostats. *Cryogenics*, **24**, 174–9, Butterworth, Guildford, UK.

Longsworth, R. C. (1986). 4 K Refrigerator and Interface for MRI Cryostats. *Advances in Cryogenic Engineering* (ed. R. Fast), **31**, Plenum, New York.

Lloyd, J. M. (1975). *Thermal Imaging Systems*. Plenum, New York.

Nisenoff, M. (1981). *Current Status of High Temperature Josephson Device Technology. Refrigeration for Cryogenic Sensors and Electronic Systems*. National Bureau of Standards, Special Publication 607, US Government Printing Office/NTIS, pp. 195–209.

Sarwinski, R. (1988). Cryogenic Requirements for Medical Instrumentation. *Advances in Cryogenic Engineering* (ed. R. Fast), **33**, Plenum, New York.

Sherman, A. (1982). History Status and Future Applications of Spaceborne Cryogenic Systems. *Advances in Cryogenic Engineering* (ed. R. Fast), **27**, 1007–11, Plenum, New York.

Sherman, A. (1983). NASA Needs and Trends in Cryogenic Cooling. In *Refrigeration for Cryogenic Sensors*, NASA Conference Publication 2287 pp. 13–28.

Voth, R. O. (1974). *Cryogenic Refrigerators for Shipboard Forward Looking*

Infrared Applications. NBSIR 74–372, National Bureau of Standards, Boulder, Colo.

Wolfe, W. L. and Zissis, G. (eds) (1978). *The Infrared Handbook*. Infra Red Information and Analysis Center, Environmental Research Institute of Michigan, Ann Arbor, Mich.

3

OPEN-CYCLE COOLING SYSTEMS

3.1 Introduction

A widely used method for low-capacity cryogenic refrigeration involves the use of a stored, cold, cryogen which eventually vaporizes and is vented to the atmosphere. The cryogen is said to be *expendable*.

The principal method is a liquid cryogen bath contained in a highly insulated dewar flask. The cold liquid vaporizes and escapes from the storage dewar. The dewar may be open to the atmosphere or sealed with a vent valve so that it operates under pressure.

To improve the cold-storage capacity the liquid cryogens may be further cooled in the storage dewar to the solid, frozen, state. In use they first melt to liquid then evaporate to vapour or alternatively they may sublime at low pressure directly to vapour from the solid state.

In yet other systems the cold cryogen may be contained in the gaseous state under a pressure greater than the critical pressure. This is called supercritical storage and is particularly suitable for space applications. The pressurized fluid has a high density and is liquid-like yet has a single phase so the fluid management problems of liquid/vapour systems do not arise with supercritical storage.

Another form of open-cycle system with expendable cryogens involves the storage of gases under very high pressure but at temperatures greater than the cryogenic range, usually ambient temperature. When cooling is required the gases are expanded in a Joule–Thomson expansion process to reduce the temperature, possibly even to change some of the gas to the liquid state. The cold fluid absorbs heat from the device to be refrigerated and is then vented to the atmosphere.

3.2 Advantages of stored cryogen cooling

Cooling by means of a liquid bath has many advantages. The greatest is perhaps the reliability of absolutely guaranteed cooling for a predictable period. It is also a relatively low-cost method of cooling, particularly for occasional use. Many storage dewars are comparatively uncomplicated and the technology for their manufacture is well known and highly developed. Considerable efforts have been invested to optimize dewars for liquid helium, in particular, with gratifying results. The initial cost of many simple types of storage dewar is surprisingly low.

Of course, it is necessary to replenish the cryogen periodically. However, in most parts of North America, the acquisition of the standard cryogens is very simple; a matter of telephoning a local supplier a short time ahead to order a particular cryogen to be delivered at a particular time and in a specific quantity. The cost is not excessive. Liquid nitrogen for example is in many cases cheaper than milk.

Cryogenic liquid baths are mechanically quiet and introduce no electromechanical effects to affect the device being cooled. They are thermally stable, even with comparatively large and intermittent or variable heat loads. Moreover they are not subject to mechanical breakdown or unscheduled interruption of the power supply.

Cryogenic liquids stored in equilibrium with their vapours can provide a convenient, constant-temperature control system. Liquids are available which provide temperatures in the range from 4.2 K to 240 K.

Solid cryogen coolers are particularly well suited for space applications where they utilize the very low atmospheric pressure of space to maintain a low vapour pressure over the subliming solid. The advantages of solid coolers over liquid systems include the higher heat content per unit mass and volume and elimination of liquid control problems due to conditions of weightlessness.

Liquid or solid cryogens provide a convenient reliable refrigeration system for low power heat sources for a surprising length of time, up to five years or even longer, depending only on the weight and volume limitations.

3.3 Disadvantages of stored cryogen cooling

The principal disadvantage of stored cryogen cooling is the inevitable loss of the cryogen due to heat leaks, principally from the environment. Periodic replenishment is therefore necessary. Moreover, some fluids, particularly helium, are not widely available in all parts of the world.

In some instances the vaporized helium is captured in inflatable plastic balloons as it vents from the dewar for subsequent reliquefaction. The provision of an on-site helium liquefier is, of course, a considerable investment and its operation and maintenance an added expense.

Even in those areas where the helium is readily available at comparatively low cost the user must still bear the cost of purchasing fresh supplies of helium and, more importantly, have available the services of a skilled technician who can effect the transfer from a storage dewar with comparatively little transfer loss. The services of such a highly skilled person are likely to be in demand for other work.

The present situation in North America, where helium is relatively cheap, readily available, and skilled technicians are not too difficult to find, may be a fools paradise. With the introduction of medical imaging systems and SQUID devices on a large scale it is inevitable that liquid helium costs will rise. Supplies will be limited and the services of skilled technicians increasingly costly and hard to find.

The continued loss of cryogen from stored cryogen systems requires the provision of relatively large volume and mass storage requirements for extended operation. This is particularly severe in situations where access for replenishment is limited or impossible, such as in space probes, hazardous environments, and so on.

The size and weight requirements may introduce limitations to gimballed, swivelling systems which may be required to operate in any orientation and to move quickly and precisely. This is not compatible with direct cooling by liquids in large heavy dewars so a flexible transfer line with rotating joints may be necessary to connect the detector to the dewar. Such transfer lines are complicated, quite expensive, and tend to have relatively high heat losses. Thus many of the advantageous features of stored cryogen systems are lost and the alternative of mechanical cryocoolers becomes attractive.

In stored cryogen systems the inevitable loss of stored cryogen precludes the use of solid or liquid cryogens for long-term storage beyond the well-defined and predictable, but limited, life prescribed by the mass of cryogen installed. Extended life may be gained by the use of large heavy systems with a large mass of cryogen and, where appropriate, the provision of secondary cryogens at higher temperatures providing the means to shield the low-temperature core in a cocoon at some higher yet cryogenic temperature, such as liquid methane or nitrogen shielding liquid hydrogen or helium.

3.4 Advantages of stored gas systems

Gases at high pressures, 40 MPa (6000 lb per sq in), may be stored indefinitely in sealed steel bottles. When cooling is required the gases are expanded through recuperative heat exchangers and Joule–Thomson expansion nozzles to rapidly produce low temperatures. In many cases the temperatures are low enough for some of the gas to be liquefied.

Systems for liquefying nitrogen, air, oxygen, or argon in this way are relatively simple. For hydrogen, neon, and helium more complicated systems are necessary involving precooling of the gas before expansion.

This method of cooling is very reliable, has the merit of relative simplicity, no moving parts, quietness in operation, and is without vibration and electromechanical emissions. Moreover, the cool-down times from start-up to the cold operating temperature are amazingly short, a few seconds for the production of liquefied gas and a minute or two for the detector and its mounting platform to be cooled (by the evaporating liquid) to operating temperature.

Stored gas, Joule–Thomson liquefiers may be made in virtually any size from miniature microcoolers generating fractions of a watt to virtually any capacity required within the limitations of availability of the stored gas supply.

3.5 Disadvantages of stored gas systems

The principal disadvantage of a stored gas system is the limited refrigeration capacity and operating life available from stored gas reservoirs of reasonable size and volume. Moreover, the system is not the most efficient thermodynamically and requires the expansion of more gas per unit of cooling than other, more complicated, mechanical coolers.

For these reasons, coupled with the advantages of quiet operation with quick cool-down, the system is particularly well suited for miniature coolers providing cooling of 0.25 W or less (at 80 K) for infrared detectors in heat-seeking missiles and night vision equipment.

However, in these miniature units the Joule–Thomson expansion nozzles are so minute they rapidly become blocked by particles of condensed contaminant thus impeding the flow of the expanding gas and so disrupting the refrigeration process. The principal contaminant in the compressed gas is water vapour. When compressing the gas it is very difficult to remove all the water (and other contaminants, oil, carbon dioxide, and so on) to the extent necessary.

The provision of very highly compressed, clean gas in bottles of convenient size is tedious under laboratory conditions and a logistical nightmare under battlefield or active service conditions. Much ingenuity and advanced engineering development has been invested in the low-capacity compression and cleaning equipment required to facilitate this process.

Systems with a larger refrigeration capacity may be equipped with an integral compressor converting the system to a closed-cycle cryocooler for continuous operation using an input of electrical power to drive the compressor.

3.6 Some fundamental considerations

To really understand the relative merits of one cryogenic refrigeration system compared to another it is necessary to be aware of the low-temperature behaviour of fluids in different states; gas, vapour, liquid, and solid. The following brief review of thermodynamic fundamentals is included for the benefit of those who for one reason or another have a need for it. Many readers will find the treatment elementary and well known. They should proceed directly to the next section describing the properties of cryogenic fluids.

Figure 3.1 shows a somewhat idealized temperature–entropy (T–S) diagram for a cryogenic fluid. For the sake of simplicity we consider

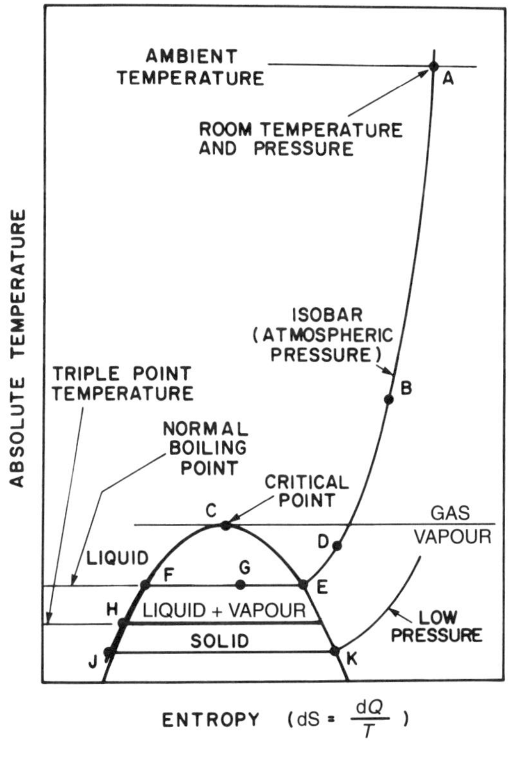

Fig. 3.1 Temperature–entropy diagram for a cryogenic fluid.

only a one-component fluid, say nitrogen or hydrogen. Mixtures of gases or liquids behave in a more complicated way.

At high temperatures, say room temperature, the fluid is in the gaseous state, point A, Fig. 3.1. At somewhat lower temperatures, say point B, the fluid remains in the gaseous state. If the cooling process takes place at constant pressure it may be represented on the temperature–entropy diagram by the constant-pressure, *isobaric*, process A–B.

On the temperature–entropy diagram, constant-pressure lines, *isobars*, are generally parallel, sloping down from the top right-hand corner to the bottom left-hand corner of the diagram. The constant pressures represented by the parallel lines increase from the right side to the left side of the diagram. Thus similar cooling processes that take place on a constant-pressure line parallel and to the left of process A–B do so at a higher pressure.

The point C, at the apex of the liquid–vapour dome, is called the *critical point* and the conditions at that point are called the critical pressure, critical temperature, critical density, and so on. The critical temperature is important for *the fluid can never exist in the liquid state at a temperature greater than the critical value*. Providing the temperature remains above the critical value, whatever is done to the gas it will always remain in the gaseous state.

When the gas at, say, state B is cooled to a lower temperature, state D, below the critical temperature, it remains in the same state but is now called, technically, a *vapour*. This is the essential difference between a gas and a vapour. One is above, and one is below the critical temperature.

Further cooling along the isobar, ABD, will reduce the temperature until, at point E, the first droplet of liquid is produced. This is called the *dew point* because the first droplet of liquid appears like the dew which forms from the water vapour in the air as the atmosphere cools overnight.

At that point, E, the fluid is said to be *dry saturated vapour*, that is, saturated with heat and completely evaporated so there is no liquid present. At any temperature above E the fluid is *superheated vapour*.

Further cooling, at a constant pressure, of the fluid at state E removes heat from the fluid but this does not cause any further decrease in the temperature. Instead more and more of the saturated vapour (at state E) is condensed to *saturated liquid* at state F. The fluid is then in two states or phases, saturated vapour at state E, and saturated liquid at state F. It is customary to represent a partially condensed mixture of saturated vapour and liquid by some intermediate point G which moves progressively along the constant-pressure line E–F as more and more of the fluid changes from saturated vapour E, to saturated liquid F. Eventually all of the fluid is condensed to saturated liquid, and point G has moved to point F. This is called the *bubble point* for it is the point when a liquid is being heated where the first bubble of vapour appears. At other pressures similar isobars to AEF may be drawn. At higher pressures the lines would be above and to the left of AEF. At lower pressures the isobars would be below and to the right of AEF.

If all the dew points and bubble points for different pressures are connected together the vapour–liquid dome shown on Fig. 3.1 is drawn. This is the locus of the various dew points and bubble points at different pressure with the *critical point* at the apex of the dome.

Within the vapour–liquid dome the fluid can exist only as saturated liquid (state F) or saturated vapour (state E). Mixtures of the fluid at these two states are, however, often represented, at point G for example, as though it were possible for a homogeneous fluid to exist at that middling state.

The *dryness fraction* of a liquid–vapour mixture refers to the fraction of the total mass that is dry, saturated vapour, state E, the remainder being saturated liquid, state F. At state G the *dryness fraction* x_G is the ratio of the length of the lines F–G and F–E, that is,

$$x_G = \frac{S_G - S_F}{S_E - S_F}$$

where S is the entropy at the point denoted by the subscript.

The heat transferred from the fluid, the enthalpy change, in condensing from state E to state F (or, conversely, in evaporating from F to E) is called the *latent heat of vaporization*.

Further cooling of the liquid results in a decreasing temperature along the isobar from F to J. In this liquid region the different constant-pressure lines are parallel but close together. At some low pressure that is characteristic of the fluid, point H, called the triple point, some of the fluid freezes and becomes solid.

At the triple point, also called the *melting point*, temperature the material exists in three states, vapour (gas), liquid, and solid. Below the melting point temperature all the liquid is frozen to the solid state. Melting ice is a common example of a material in the three states of matter: vapour, liquid, and solid.

At temperatures less than the critical temperature the pressure of the vapour above the liquid or solid material, called the *vapour pressure*, depends only on the temperature in a fixed relationship that is characteristic of the element.

If the pressure of the nitrogen vapour above liquid nitrogen is 1 atmosphere (14.7 lb per sq in) the temperature of the nitrogen liquid and vapour is 77 K—it can be nothing else. At lower temperatures the vapour pressure is lower. At higher temperatures the vapour pressure is higher.

The temperature of the liquid when the pressure (of the liquid and vapour) is 1 atmosphere is called the *normal boiling point*.

When the frozen fluid, a solid, has a temperature J, say, less than the melting point temperature, the vapour pressure is relatively low and has an equilibrium value which depends only on the temperature. It depends on the particular fluid whether or not the pressure is greater or less than 1 atmosphere.

If now the frozen solid at state J is heated and the pressure above the solid is maintained at the same low value the frozen solid will not melt. Instead it releases vapour directly from the solid. The vapour is said to *sublime* from the solid.

The enthalpy change from frozen solid at state J to dry saturated vapour at state K is the *latent heat of sublimation* of the solid. The latent heat of sublimation is, in fact, the change in entropy $(S_K - S_J)$ times the temperature at which the sublimation occurs.

Since the change in entropy is defined as the heat transferred divided by the temperature at which the change occurs, then $dS = dQ/T$. Rearranging the terms $dQ = T\,dS$. Now the real value of the temperature–entropy diagram is revealed. It provides a very simple way for estimation of the heat transferred when a fluid or solid changes its condition from one state to another. If the end states are known and the nature of the process connecting them is known, such as constant pressure, then the area of the T–S diagram under that process is the heat transferred in that process.

Turning now to Fig. 3.2, this is basically a reproduction of Fig. 3.1 with the additional constant-pressure cooling process LMN at pressure p_2. It can be seen from Fig. 3.2 that the isobar p_2 is to the left of the critical pressure isobar, p_c, shown as the broken line.

Gas at the higher pressure, p_2, is said to be at a *supercritical pressure*. The significance of this is that, *no matter how low the temperature is reduced, the compressed fluid never becomes liquid*. It always remains a dense, cold, homogeneous fluid at low temperatures, more akin to liquid than gas but nevertheless not experiencing the normal phase change of that same fluid when cooled at subcritical pressures.

Supercritical cryogenic storage is important principally in space operations. In the low gravity of the space environment the vapour and liquid behave differently from a normal gravity situation. There are no strong gravity forces and so surface tension or capillary forces, normally

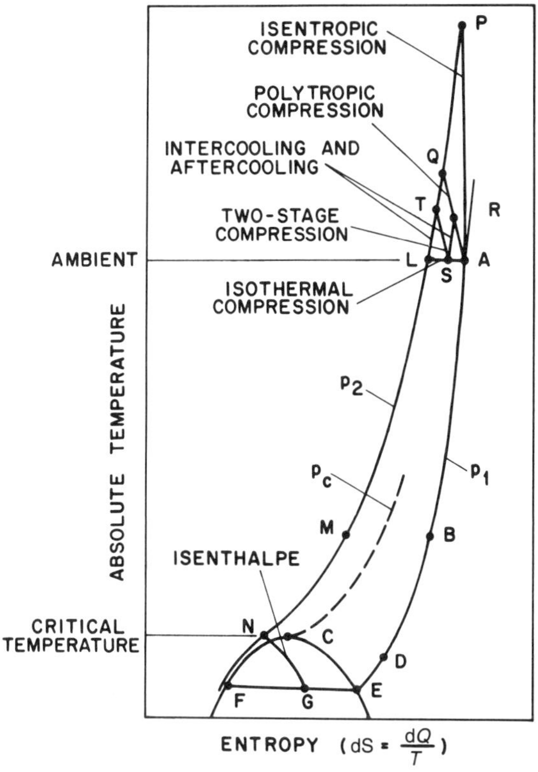

Fig. 3.2 Temperature–entropy diagram for a cryogenic fluid.

insignificant, need careful consideration in the fluid management scheme. The homogeneous, superfluid cryogen is not affected by gravity or surface tension forces for there is no separate liquid and vapour fluid.

It can be seen from the above brief introductory review that the temperature–entropy diagram for a given fluid is an extraordinarily useful instrument for thermodynamic analysis and computing the heat transferred in different processes.

In addition to those processes discussed above many other thermodynamic processes can be displayed on the temperature–entropy diagram. For example, referring again to Fig. 3.2 the process N–G is the constant-enthalpy or *isenthalpic* process experienced in the Joule–Thomson expansion of a compressed gas from a pressure p_2 at state M to the low pressure p_1. The final condition of the fluid, at state G, indicates some saturated liquid, state F, is produced, the remainder being dry saturated gas, state E.

The process A–L on Fig. 3.2 is the constant-temperature *isothermal* compression process from pressure p_1 to pressure p_2 at ambient temperature. The process A–P is the constant-entropy *isentropic* compression process from pressure p_1 to p_2.

The process A–Q is the more realistic *polytropic* compression process p_1 to p_2. This may take place in the single stage $A-Q$ shown, but more often in several stages A–R (first stage) and S–T (second stage) with water-cooled *intercooling*, R–S, between stages and water-cooled *after-cooling*, T–L after compression.

When heat is transferred at constant pressure from the compressed gas the process may be presented as the cooling process L–N. This heat may be imparted to the low-pressure gas reheating from state E to state A. The heat exchange process occurs in a *recuperative heat exchanger*.

The refrigeration *load* is the enthalpy change of the fluid from state G to state E, basically the vaporization of that fraction of fluid that is saturated liquid at state F to the dry saturated vapour state E. Heat is drawn from the surrounds of the container holding the cryogenic liquid. The container thus becomes cold, tending to assume the temperature of the vaporizing liquid.

3.6.1 Properties of cryogenic fluids

A very important contribution of the now defunct Cryogenics Division of the US National Bureau of Standards (NBS)† at Boulder, Colorado, was the compilation of tables of thermodynamic and transport properties of the cryogenic fluids and the preparation of the associated temperature–entropy and other thermodynamic property charts. A useful compendium of the temperature–entropy charts for the customary cryogenic fluids was given by Barron (1984) and by Timmerhaus and Flynn (1989). They are too small for general use but the quality of the printwork is excellent and photographic or xerographic enlargement will provide a working diagram.

Tables of the thermodynamic and transport properties of the customary cryogenic fluids were also included in both books, the recommended foundation volumes for any cryogenic library or bookshelf.

Radebaugh (1993) reviews the sources of available material properties including the NBS publications and most importantly the computer

† Recently transformed to the National Institute of Science and Technology (NIST).

codes for fluid properties (tapes or floppy disks) available from the NBS Office of Standard Reference Data. He also reviews the solid materials reference sources.

CRYODATA (1993) of Niwot, Colorado, has available at modest cost two computer disk. One, CPPAK, is concerned with specific heats and enthalpies of technical solids at low temperatures. Properties for thirty-seven different solids, metals, and non-metals are included. The other disk HEPAK is concerned with calculating the properties of helium including the usual thermodynamic and thermophysical properties over a very wide range of temperatures and pressures.

Both the CRYODATA programs are essentially based on earlier NBS publications prepared by the owners prior to their retirement.

Another excellent compilation of thermodynamic data and thermodynamic charts for refrigerant fluids, including the cryogens, will be found in the *Handbook of Fundamentals* published by the American Society of Heating, Refrigerating and Air Conditioning Engineers (ASHRAE), New York.

Table 3.1, reproduced from Donabedian (1978), summarizes the important physical properties of some of the fluids with normal boiling points less than ambient temperature. This table is a most useful compendium of the properties of fluids that are important in stored cryogen bath refrigeration.

In the table the fluids are generally arranged in increasing order of the normal boiling point. Column 1 shows the name of the fluid and its chemical description. Column 2 shows the normal boiling point (the temperature of the liquid at which the vapour pressure is 1 atm). Column 3 shows the melting or triple point (temperature) at 1 atmosphere pressure. Columns 4, 5, and 6 show the densities of (a) liquid at the normal boiling point, (b) gas at 273 K at 1 atm pressure, and (c) the vapour at normal boiling point (by definition 1 atm pressure).

Column 7 shows the fluid vapour pressure at the melting (triple) point. Columns 8 and 9 show the heat of vaporization and heat of fusion, respectively, of the subliming, melting, or evaporating fluid.

Finally columns 10, 11, and 12 show the critical parameters, temperature, pressure, and volume.

Another useful chart given by Donabedian (1978) is reproduced in Fig. 3.3. This shows the operating temperature range of selected expendable coolants including both the solid phase (sublimation cooling) and liquid phase (evaporative cooling). For all fluids the range of operation is limited to a minimum pressure of 0.1 mm Hg vapour pressure, generally accepted as the minimum practicable venting pressure to space, through to the critical pressure (usually far too high for operation at these pressures). A more practical cut-off maximum pressure for

Table 3.1 Properties of miscellaneous liquids, gases, refrigerants, and fuels (after Donabedian 1978).

	1	2 Boiling point (bp) at 1 atm pressure (K)	3 Melting point (mp) at 1 atm pressure (K)	4 Liquid density at bp ($g\ ml^{-1}$)	5 Gas density at 273 K ($g\ l^{-1}$)
Helium	He^3	3.2	~1.0†	–	–
Helium	He^4	4.2	~2.0††	0.125	0.1735
Hydrogen	H_2	20.39	13.98	0.071	0.0899
Deuterium	D_2**	23.6	18.7	0.173	0.167
Tritium	T_2**	25.1	21.6	–	–
Neon	Ne	27.2	24.47	1.2	0.901
Nitrogen	N_2	77.37	63.4	0.808	1.250
Carbon monoxide	CO	81.6	68.0	0.812	1.186
Fluorine	F_2	85.24	53.6	1.513	1.71
Argon	A	87.4	83.6	1.391	1.78
Oxygen	O_2	90.1	54.9	1.14	1.43
Methane	CH_4	111.7	90.7	0.425	0.72
Krypton	Kr	120.3	116.0	2.4	3.75
Freon-14	CF_4	145.14	89.5	1.62	–
Ozone	O	161.3	80.5	1.46	2.14
Xenon	Xe	165.3	150.5	3.1	5.93
Ethylene	C_2H_4	169.3	104.0	0.578	1.19
Nitrous oxide	N_2O	183.6	183.0	1.23	1.84
Ethane	C_2H_6	184.8	90.0	0.562	1.28
Acetylene	C_2H_2	189.1	191.2	0.623	1.09
Freon-13	$CCLF_3$	192.0	91.6	1.505	–
Carbon dioxide	CO_2	194.6	215.7	1.51	1.87
Propylene	C_3H_6	226.1	77.5	0.604	1.78
Freon-22	$CHCLF_2$	232.5	113.0	1.414	21.3
Ammonia	NH_3	239.8	195.0	0.683	0.77
Freon-12	CCL_2F_2	243.1	118.0	1.488	17.7
Methyl chloride	CH_3Cl	249.4	–	0.993	5.93
Sulfur dioxide	SO_2	263.1	198.0	0.80	4.49
Isobutane	C_4H_{10}	272.5	–	–	2.53
Freon-11	CCL_3F_2	296.8	162.7	1.48	2.47
Propane	C_3H_8	230.8	85.9	0.595	1.92

* The critical volume is the volume at the critical temperature and critical pressure.
** Isotopes of hydrogen.
† At 25 atm.
†† At 29 atm.

6 Vapour density at bp (g ml^{-1})	7 Vapour pressure solid at mp (mm Hg)	8 Heat of vapour at bp (J g^{-1})	9 Heat of fusion at at mp (J g^{-1})	10 Critical temperatures (K)	11 Critical pressure (atm)	12 Critical volume* (l g^{-1})
–	–	–	–	–	–	–
17.0	–	20.5	4.183	5.2	2.26	0.0144
1.286	54.0	448.0	58.15	33.19	12.98	0.03321
2.58	12.8	286.0	50.0	38.3	16.2	0.0142
–	188.0	–	–	43.7	20.8	0.0089
9.5	323.0	87.0	16.72	44.5	26.8	0.002
4.415	96.5	199.0	25.52	126.1	33.5	0.00321
–	–	213.5	29.27	138.8	35.0	0.0032
–	0.1	171.5	13.4	144.8	55.0	–
5.03	516.0	162.7	28.05	150.8	48.0	0.0019
4.75	2.0	212.5	13.8	154.1	50.1	0.0023
1.76	71.0	581.0	60.25	190.5	45.8	0.008
8.33	550.0	108.0	16.3	209.3	54.5	–
7.2	–	134.8	–	227.5	37.0	–
–	–	316.0	13.8	261.1	54.6	0.00306
9.77	615.0	96.25	119.1	290.0	58.0	0.00086
2.08	–	481.0	148.5	282.8	50.9	0.0045
–	858.0	250.2	95.0	399.7	71.7	0.0022
0.32	–	490.0	96.25	305.0	48.8	0.6048
–	–	916.0	–	309.0	62.0	–
7.9	–	146.4	179.9	302.0	3.9	–
–	–	574.0	71.6	304.5	73.0	0.0022
–	–	439.5	–	365.0	45.0	–
4.65	–	235.0	35.15	369.0	48.7	–
0.898	45.0	1363.0	–	405.0	111.2	0.0042
6.25	–	167.2	–	384.0	39.6	–
2.56	–	427.0	–	–	–	–
3.2	–	388.0	–	430.0	77.7	0.002
–	–	–	–	426.0	36.0	–
5.93	–	237.3	–	471.0	43.1	–
2.08	–	342.0	79.9	370.0	42.0	–

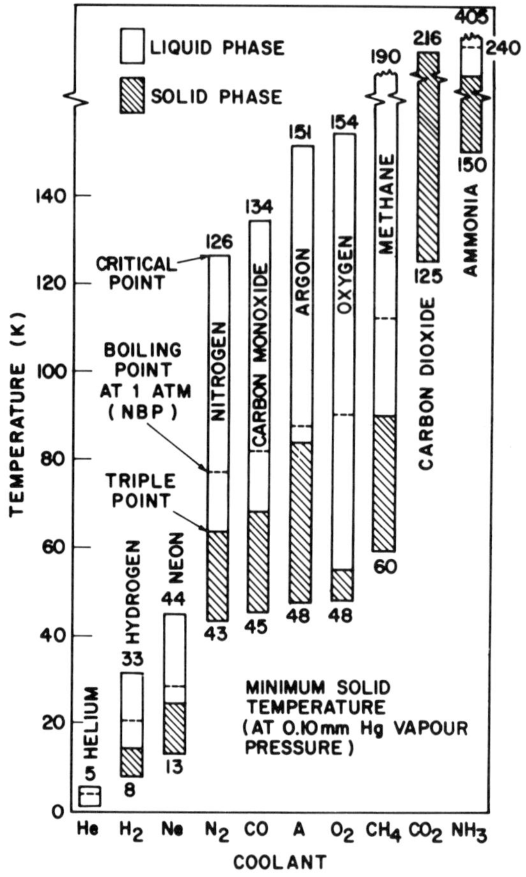

Fig. 3.3 Operating temperature range achievable with stored cryogenics (after Donabedian 1978).

actual operation might be 5 or even 10 atm in order to keep the pressurized, insulated storage dewar at reasonable weight (and cost).

Figure 3.3 shows that cooling by stored cryogens can be provided from about 2 K to over 300 K by careful selection of the coolant under the proper environmental conditions. Figure 3.4 displays similar data for selected cryogens on a pressure–temperature plane (note the pressure scale is logarithmic) from a low vapour pressure and temperature of 10^2 Pa† and 10 K, respectively, to the critical points of the various fluids.

† Since 1 atm = 15 lb per sq in approx. = 10^5 Pa approx. = 860 mm Hg, therefore 10^2 Pa = $10^2/10^5 \times 860$ = 0.86 mm Hg = 0.015 lb per sq in.

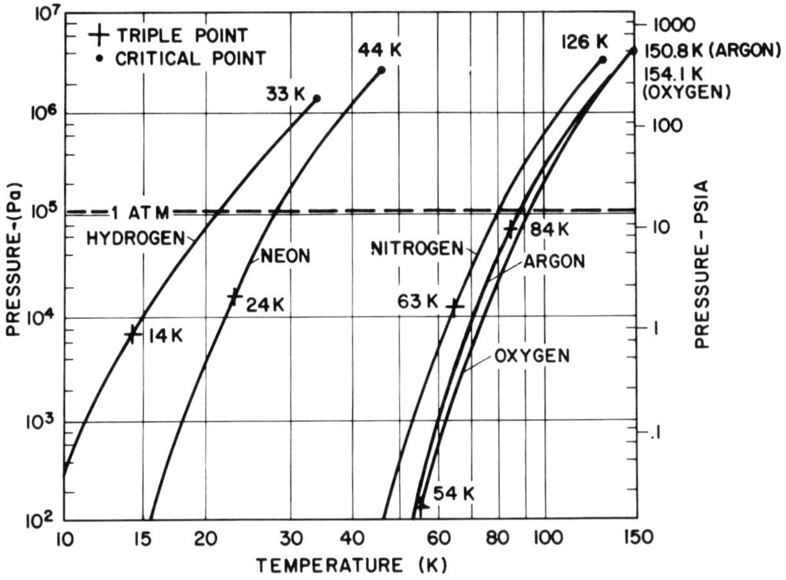

Fig. 3.4 Pressure–temperature relationship for solid and liquid cryogens.

3.7 Storage dewars: some aspects of design

The storage containers for liquid and solid cryogens are, essentially, metal or glass tanks that are very highly insulated. Unless there is some compelling reason the tanks are invariably cylindrical or spherical in form so as to enclose the maximum volume with the minimum surface area (to minimize the heat gain, weight, material used, and cost).

The primary insulation is usually a vacuum jacket enclosing the whole tank with perhaps additional urethane foam insulation to provide some protection against shock and impact as well as thermal protection and a protective metal jacket equipped with lifting, securing, or transporting devices.

A high vacuum (10^{-5} torr or better)† is developed in the insulating space during manufacture. In many cases a *getter* material is included in the vacuum space so as to maintain the vacuum in service.

The purpose of the vacuum space is, of course, to thermally isolate the cold cryogenic liquid from convective heat transfer effects coupling

† 1 torr = 1 mm Hg pressure.

the cold interior space to the outer case at ambient temperature. To further reduce the thermal coupling the internal surfaces of the vacuum insulation are coated, by plating or vapour deposition with gold, mercury, or silver. These highly polished, highly reflective surfaces have a low thermal emittance or absorbtivity (that is, $\epsilon = 0.04$) so the radiation heat transfer between the inner and outer shells is minimized.

Heat leaks from the environment at room temperature are further minimized by the interposition of radiation shields between the inner and outer shell. The inner radiation shield may be thermally coupled to the outlet vent of the internal flask. The boil-off vapour escaping through the vent thus cools the inner shield which is then described as a *vapour-cooled shield*. In a liquid helium dewar the vapour-cooled shield frequently operates at 30–40 K so the radiation heat transfer is reduced to a minimum.

Further reductions are achieved with the addition of more shields. This is taken to the limit with the addition of a *superinsulation multilayered blanket* of alternate layers (200 or more) of thin aluminized Mylar sheets separated by thin cotton or nylon net. This multilayer insulation presents multiple radiation shields to minimize the radiation heat transfer and it is most effective when incorporated in a highly evacuated space.

In many instances the very low-temperature cryogens (helium, hydrogen, neon, and so on) can be thermally protected by an outer jacket of higher temperature, lower cost, and sacrificial fluid, nearly always nitrogen boiling at 77 K. This in turn may be thermally isolated within a high-vacuum insulating space silvered and equipped with shields as described above.

There must be an inlet port and an outlet vent to fill and exhaust the internal flask. In many cases, too, there is an open neck by which the object or device to be cooled may be introduced to the cryogen bath contained by the inner flask. The inner flask must be mechanically supported in some way which generally involves some connection to the outside container.

For all these connections careful design is necessary to minimize conduction heat transfer from the warm outside to the cold inner system. Low-conductivity materials, stainless steel, epoxy, fibreglass, and special glasses with a low coefficient of thermal conductivity are used for these conduction couplings. Conduction heat transfer is minimized if long, thin section elements in tension are used rather than short stubby columns in compression.

The device being cooled may be electrically active, such as a superconducting magnet, so there will need to be power leads to and from the device and perhaps other fine leads for data acquisition, control, and the like.

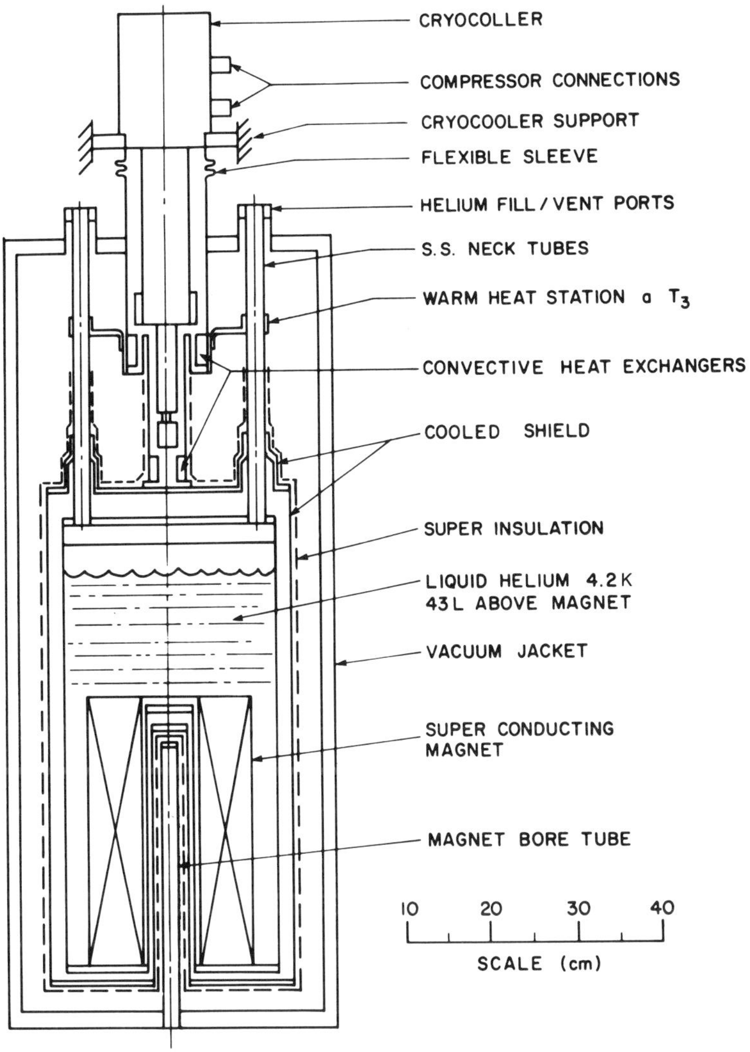

Fig. 3.5 Helium storage dewar (after Longsworth 1978).

Heat conducted along these leads is often an important and significant component of the total heat leak from the outside. The lead heat leak will be minimized by the use of long thin wires having the minimum acceptable cross-section. Of course this can be carried too far if the result is an electrical I^2R resistance loss that is significant compared with the thermal conduction leak.

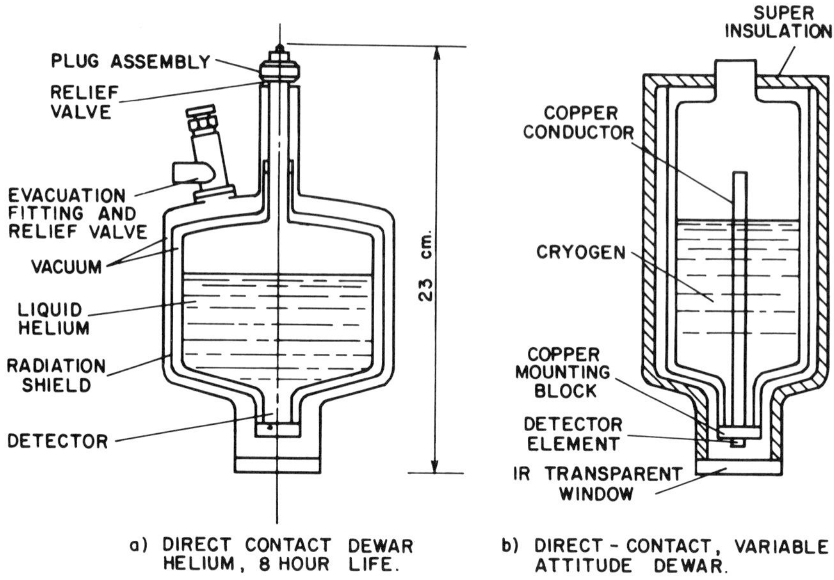

Fig. 3.6 Direct contact storage dewars (after Donabedian 1978).

In non-space applications where normal gravity prevails it is customary, wherever possible, to operate the storage dewar in the vertical orientation with the coldest parts at the bottom. This minimizes any convective heat transfer action that may remain in the system. The coldest liquid or vapour will fall, under gravity, to the lower-most part of its possible travel.

Finally it is advisable to encapsulate the system in a protective (mechanical shock, impact, and vibration as well as thermal) layer of urethane foam and an outer skin of metal or plastic complete with lifting handles, locking caps, robust base, and all the necessary identifying and warning notices and displays.

A helium dewar that incorporates many of these design principles is shown in Fig. 3.5. A superconducting magnet is contained in the helium bath with a magnet bore tube accessible from below the dewar to insert and withdraw experimental devices without penetrating the dewar.

The dewar is equipped with vapour-cooled shields and also includes the two-stage expansion element of a Gifford–McMahon cryocooler to provide supplementary cryogen cooling at 80 K and 20 K to reduce the liquid helium boil-off rate.

Smaller storage dewars with an infrared or other type of detector built directly into the dewar are shown in Fig. 3.6. These are said to be *direct contact dewars*. Figure 3.6(b) shows the detector mounting block equipped with copper conducting paths to permit operation in any orientation yet maintain good thermal connection between the detector and the cryogen coolant.

3.8 Solid cryogen dewars

Storage dewars for solid cryogen storage incorporate the same fundamental design principles as liquid cryogen storage dewars. Vapour sublimes preferentially from the solid cryogen at the point of thermal input so it is necessary to provide a thermal conduction path between the solid cryogen and the detector mount or other device to be cooled.

For this purpose, various types of porous expanded metal foams and wire mesh heat exchangers have been used to couple the solid cryogen to solid metal conduction rods coupled to the detector base. Sometimes cryogenic heat pipes have been used.

The advantages of a solid cryogen system compared with a liquid cryogen are that lower temperatures are possible, permitting an improvement in sensitivity for some infrared detectors and other cooled devices. Moreover, the solid has a higher storage capacity since the heat of sublimation is the sum of the heat of vaporization and fusion. Therefore a higher density of cold storage is possible. Furthermore, a solid system can operate in any orientation without concern for gravitational effects.

Sherman (1982) briefly described the CLAES† solid hydrogen cooler incorporating a storage dewar shown in Fig. 3.7 for cooling 23 Si (Ge) detectors and the spectrometer optics to 12 K with vapour cooling of the telescope baffle to 130 K.

The cooling load at 12 K was specified by Sherman (1982) to be 0.3 W with the baffle load of 2.4 W and an eighteen month life requirement. The anticipated weight of the system was 354 kg including 78 kg of hydrogen.

Solid cryogen cooling at higher temperatures with multiple levels of cooling is also possible. Figure 3.8 shows a two-stage, one-year lifetime, solid cryogen cooler utilizing solid methane (CH_4) at 60K and ammonia (NH_3) at 150 K built by the Lockheed Space and Missile Company,

† CLAES: Cryogenic Limb Array Etalon Spectrometer instrument on the Upper Atmospheric Research Satellite (UARS).

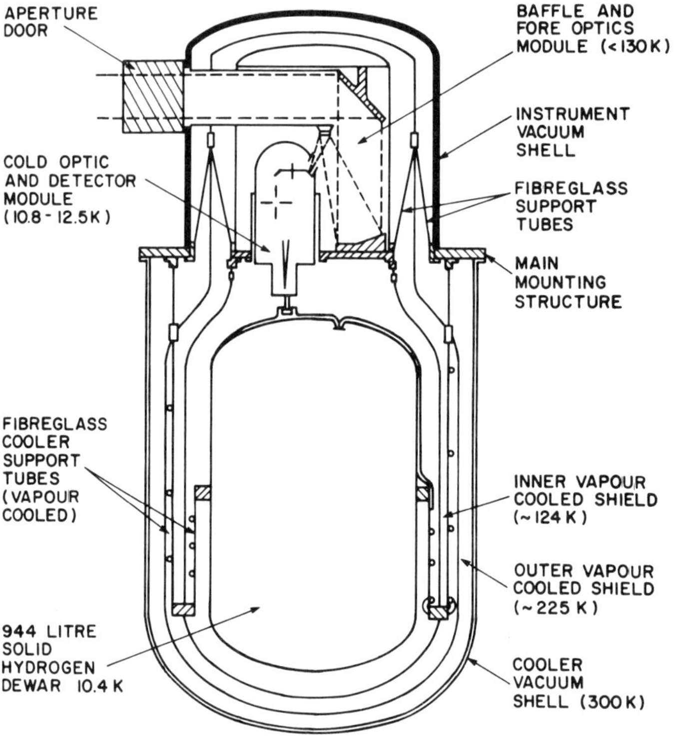

Fig. 3.7 CLAES solid hydrogen cooler (after Sherman 1982).

Sunnyvale, California, for NASA as part of the Limb Radiance Inversion Experiment and launched in June 1975 on the Nimbus-F spacecraft (after Donabedian 1978).

3.9 Solidification processes

Cryogens used in the solid state are usually installed in the storage dewar as liquids and solidified *in situ*. The cryogenic liquid is usually cooled to the triple point temperature and below by circulating a cold fluid, such as liquid helium at 4.2 K, to cool and solidify hydrogen from 20 to 10 K, through cooling coils provided in the storage dewar.

Alternatively, cooling of the cryogenic liquid may be effected by decreasing the vapour pressure above the liquid by vacuum pumping the vapour away. As the vapour pressure is reduced below the equilibrium

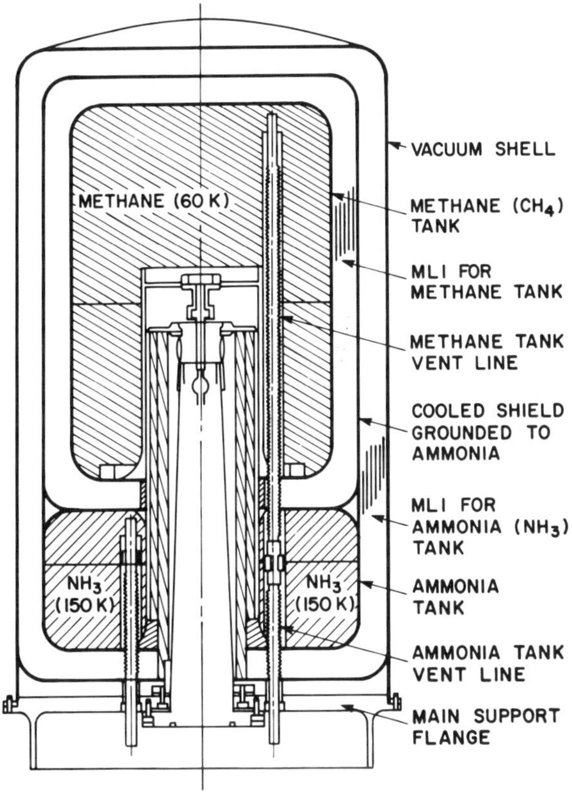

Fig. 3.8 Two-stage, one-year, methane/ammonia solid cryogen cooler flown on the Nimbus-F spacecraft (after Donabedian 1978).

vapour pressure more vapour is evaporated from the liquid to restore the pressure to its equilibrium value. The heat to evaporate the replacement vapour is drawn from the body of the liquid which cools in consequence. This process continues to the triple point temperature when solidification occurs and then further to subcool the solid.

3.10 Supercritical storage of cryogens

The storage of liquid cryogens in space presents problems in fluid management. In the absence of gravity or other acceleration forces the effects of surface tension and other minor forces, negligible in earth-bound systems, become significant. Random orientation of the liquid

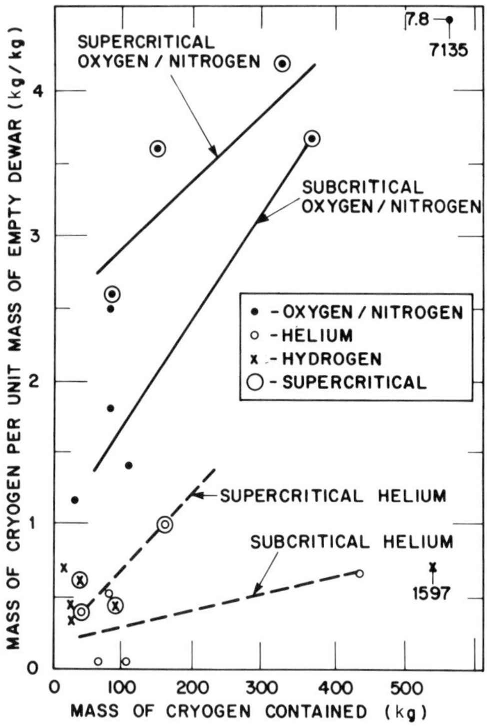

Fig. 3.9 Mass of cryogen per unit mass of empty dewar as a function of the mass of cryogen contained (after Donabedian 1978).

phase during weightless conditions prevents continual contact between the liquid and detector or supply point. Moreover, it is difficult to predict liquid behaviour unless 'wick' systems are provided to bring capillary surface tension forces to bear upon the errant liquids.

Another solution is to pressurize the storage dewar to a pressure greater than the critical pressure. Under these conditions there can be no liquid present. The fluid is a dense, homogeneous, liquid-like fluid but without the separation problems of a liquid–vapour two-phase system.

The critical pressure of helium is only 2.26 atm (Table 3.1) whereas for hydrogen the critical pressure is 13 atm, for oxygen 50 atm, and for nitrogen 33.5 atm. The relatively high pressures of supercritical storage require stronger containers than subcritical dewars where the pressure rarely exceeds one or two atmospheres.

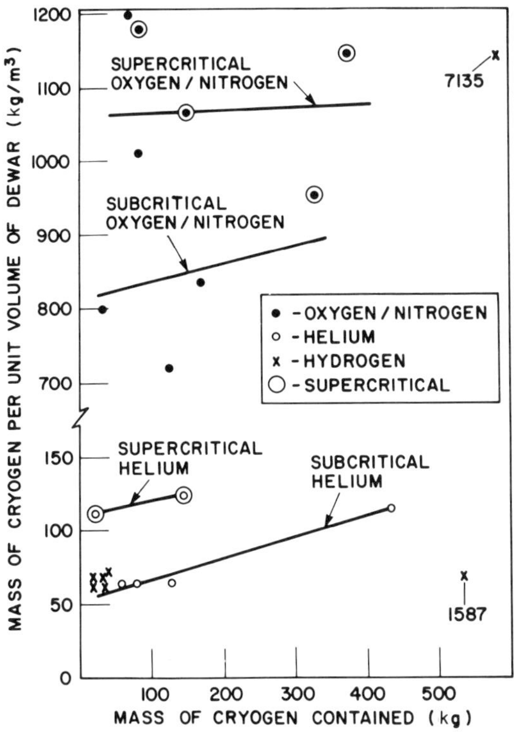

Fig. 3.10 Mass of cryogen per unit volume of dewar as a function of cryogen contained.

Donabedian (1978) listed the characteristics of 24 large high-performance cryogenic fluid dewars designed for airborne and space systems. Some of these were supercritical storage and some were sub-critical containers.

These characteristics were used to prepare Fig. 3.9 showing the mass of cryogen per unit mass of (empty) storage dewar as a function of the maximum mass of cryogen contained by the dewars. Data was included for oxygen and nitrogen dewars, for hydrogen dewars, and for helium dewars.

As might be expected the mass of cryogen per unit mass of dewar was greatest for the oxygen and nitrogen systems. There was little difference between the data for the hydrogen and helium dewars. For all the fluids the mass of cryogen contained per unit mass of dewar increased with size (that is, payload of cryogen) although there were wide variations.

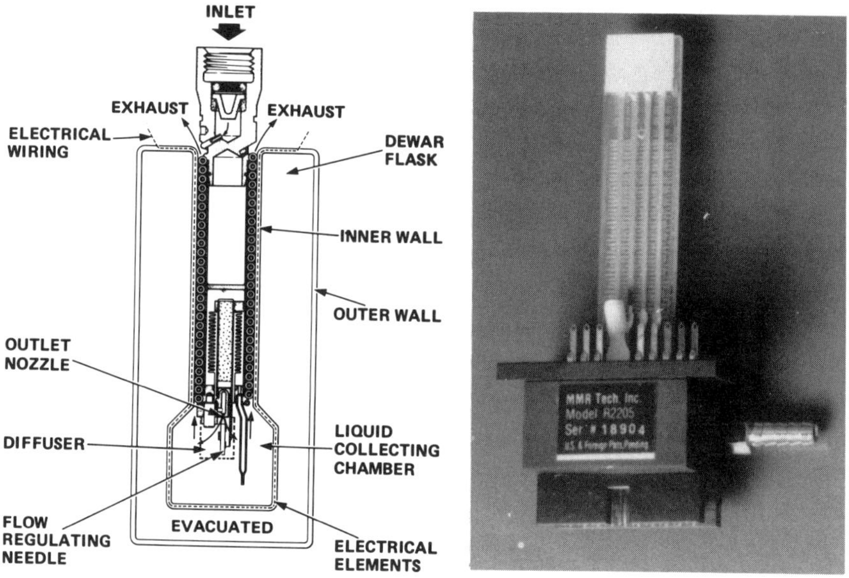

(a) (b)

Fig. 3.11 Miniature Joule–Thomson expansion units; (a) courtesy Hymatic Engineering Co. Ltd., (b) courtesy MMR Technologies Inc.

An unexpected result of displaying the data in this way was that the supercritical storage systems were found to be as good and often better, in terms of the mass of cryogen per unit mass of dewar, than the subcritical liquid storage systems. It was anticipated intuitively that the supercritical storage systems would be heavier than the subcritical storage systems.

Insufficient data on these systems is available to satisfactorily explain this result. Perhaps the explanation lies in the fact that cold fluids at supercritical pressures have a greater density than the liquids and therefore a given mass occupies a smaller volume. The savings in weight of the smaller container may offset the thicker wall necessary to contain the higher pressures.

Confirmation of this hypothesis was obtained with the data displayed in Fig. 3.10 showing the mass of cryogen contained per unit volume of the dewar as a function of the mass of the cryogen. This data was drawn from the same table of data given by Donabedian (1978) for

24 large high-performance cryogenic dewars designed for airborne and space systems. It shows the supercritical storage dewars were superior to the subcritical storage dewars in terms of the mass of cryogen stored per unit volume of the dewar.

3.11 Stored gas systems

The miniature cryocoolers most widely used at the present time are shown in Fig. 3.11. They employ Joule–Thomson or isenthalpic expansion of gas compressed to very high pressure to achieve the desired cooling and are contained or encapsulated in a vacuum-insulated glass or metal dewar flask. The expanded gas is cooled sufficiently by the expansion to be partially liquefied and this is used to cool the infrared detector or other electronic element located in the cold region.

Figure 3.11(a) shows a miniature Joule–Thomson expansion unit with coil-tube Giaque–Hampson heat exchanger in an insulated glass dewar vessel. They are used in very large numbers for night vision and heat-seeking missile guidance systems. Figure 3.11(b) shows one of the new planar microminiature Joule–Thomson expansion units fabricated using a photolithographic technique. Used presently for a variety of laboratory and instrument applications these units are expected to find very large scale application in microminiature cold electronics systems.

3.11.1 Joule–Thomson expansion

A Joule–Thomson or isenthalpic† expansion occurs when a gas at high pressure expands to a lower pressure and does so with no significant change in gas velocity, with no work produced, and no heat transferred. Instead the gas expands through a restricted orifice or long narrow channel in such a way as to create high fluid friction thereby dissipating the potential for useful work output inherent in the initial high pressure. With high fluid friction the expansion process is thermodynamically irreversible, that is, the expansion can only occur from a high pressure to a low pressure. (It is inconceivable that fluid at a low pressure could enter a thermodynamic 'black box' and leave at a high pressure without the expenditures of work or heat energy to accomplish the compression.)

† Enthalpy H, is a thermodynamic property of a gas. It is not a fundamental property but is the sum of the internal energy U plus the product of the pressure p and the volume V, that is, $H = U + pV$. For an ideal gas (one that obeys the ideal gas law $pV = mRT$) the temperature in an isenthalpic expansion remains constant.

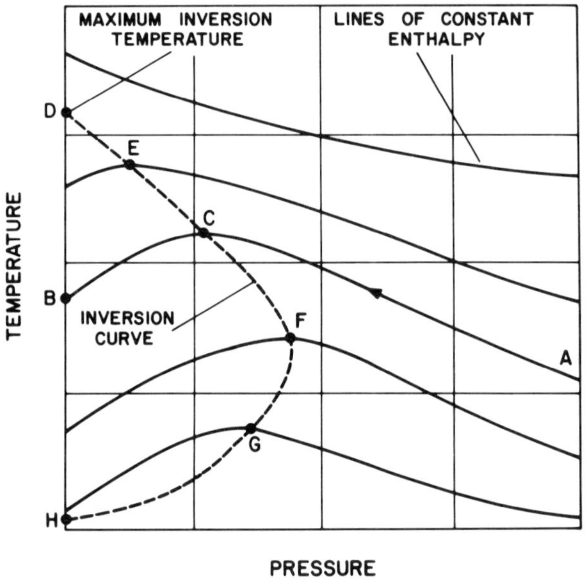

Fig. 3.12 Generalized isenthalpes for a non-perfect gas drawn on the pressure–temperature plane.

The Joule–Thomson expansion process is very common. All domestic and commercial refrigerators operating on the vapour-compression cycle incorporate a Joule–Thomson expansion valve. Another common example is the gas escaping through a control valve that is partly cracked open so the flow area is very small allowing only partial flow with very high friction.

In theory, a perfect gas (a gas that obeys the law $pV = mRT$) experiencing an isenthalpic expansion would have no change in temperature throughout the expansion. However, the behaviour of a real gas experiencing isenthalpic expansion is quite different. This is illustrated in Fig. 3.12 which shows generalized constant-enthalpy curves displayed on the pressure–temperature plane.

Starting at the right-hand, high-pressure side of the figure, at point A, any expansion of the gas to a lower pressure in a constant enthalpy (high-friction, porous plug-like) expansion will cause an increase in temperature. This continues until, at point C, the constant-enthalpy line achieves the highest temperature. Any further expansion will then result in a decrease in the temperature (to B at the lowest possible pressure).

Table 3.2 Maximum inversion temperature for different gases (after Barron 1985).

Gas	Maximum inversion temperature (K)
Above ambient	
Ammonia	1994
Carbon Dioxide	1500
Methane	939
Oxygen	761
Argon	794
Nitrogen	621
Air	603
Below ambient	
Neon	250
Hydrogen	205
Helium 4	40

Similar curves may be drawn for different temperatures at the start of the expansion. When all points at the apexes of the isenthalpes (corresponding to C) are joined, as shown by the broken line DCE, the resulting line is called the *inversion curve*. The upper end of the inversion curve is called the *maximum inversion temperature*.

It can now be seen that in isenthalpic expansion, cooling of the gas will only be obtained if the process occurs to the left of, or inside, the inversion curve. Starting the expansion at a higher pressure than the limit of the inversion curve (point C) serves no useful purpose for during the early part of the expansion the temperature will increase rather than decrease.

Moreover, if the temperature at the start of the expansion is so high that the isenthalpe never enters the inversion curve then, whatever the pressure, there can never be a decrease in the temperature during isenthalpic expansion. The temperature will always increase or, at best, remain constant.

Table 3.2 shows the maximum inversion temperature for some common gases. In many cases this is above the normal ambient temperature (300 K) but neon, hydrogen, and helium have maximum inversion temperatures that are less than ambient. If these gases are expanded isenthalpically from a high pressure and at ambient temperature they will always experience an increase in temperature. To achieve a decrease in temperature they must first be cooled to temperatures below the maximum inversion temperature. It is clear from Fig. 3.12 that to achieve a significant decrease in temperature on expansion, the precooling must be carried out to a temperature well below the maximum inversion temperature.

For helium the maximum inversion temperature is 40 K. Therefore, if a significant fraction (say 10 per cent) of the gas is to be liquefied the compressed helium must be precooled to 20 K or less. This precooling may be done by a closed-cycle mechanical cryocooler or perhaps by a liquid hydrogen bath, itself produced by a Joule–Thomson expansion of compressed hydrogen. Hydrogen has a maximum inversion temperature of 205 K so, to produce the liquid hydrogen, the compressed hydrogen gas must itself be precooled. Precooling of the hydrogen, and initial precooling of the helium, can be accomplished using nitrogen, oxygen, argon, or air. All these gases have inversion temperatures greater than ambient and so do not need to be precooled in order to decrease their temperatures when expanding from high to low pressure.

This type of multifluid system wherein helium is precooled by hydrogen and both are precooled by nitrogen prior to their Joule–Thomson expansion is called a cascade Joule–Thomson cooler. Triple-fluid coolers of this type are not in commercial series production but several experimental units have been constructed and are described in the literature. The technology for their manufacture is well developed and understood and efforts to develop commercial versions are currently in progress. Such machines are likely to be large and relatively complicated, involving several mechanical compressors, and expensive, so the alternative multiple-expansion mechanical cryocoolers may be more attractive commercially.

3.11.2 Joule–Thomson coefficient

The change in temperature for an isenthalpic pressure drop is defined by the Joule–Thomson coefficient for a fluid $\mu_{\mathrm{JT}} = (\partial T/\partial p)_H$, the change in temperature due to a change in pressure at constant enthalpy. The Joule–Thomson coefficient is the slope of the isenthalpic curves shown in Fig. 3.12. The coefficient is zero along the inversion curve (at the peak of the isenthalpes) since the slopes of the isenthalpes are zero. To the right of the inversion curve the Joule–Thomson coefficient is negative and the temperature increases on expansion. To the left the Joule–Thomson coefficient (the slope of the isenthalpe) is positive and the temperature decreases on expansion.

The Joule–Thomson coefficient can be written†

† The following is a summary of the excellent treatment of this topic given by Barron (1984).

$$\mu_{JT} = \left(\frac{\partial T}{\partial p}\right)_H = -\left(\frac{\partial T}{\partial H}\right)_p \left(\frac{\partial H}{\partial p}\right)_T \tag{3.1}$$

and from elementary thermodynamics

$$dH = \left(\frac{\partial H}{\partial T}\right)_p dT + \left(\frac{\partial H}{\partial p}\right)_T dp \tag{3.2}$$

and

$$dH = C_p\ dT + \left[v - T\left(\frac{\partial v}{\partial T}\right)_p\right] dp \tag{3.3}$$

where v is the specific volume of the material.

A comparison of the coefficients of dT and dp in Eqs (3.2) and (3.3) indicates that

$$C_p = \left(\frac{\partial H}{\partial T}\right)_p \tag{3.4}$$

and

$$\left(\frac{\partial H}{\partial p}\right)_T = v - T\left(\frac{\partial v}{\partial T}\right)_p . \tag{3.5}$$

Combining Eqs (3.1), (3.4), and (3–5) we may write

$$\mu_{JT} = \frac{1}{C_p}\left[T\left(\frac{\partial v}{\partial T}\right)_p - v\right] . \tag{3.6}$$

For an ideal gas, $v = RT/p$ and

$$\left(\frac{\partial v}{\partial T}\right)_p = \frac{R}{p} = \frac{v}{T} \tag{3.7}$$

so that from Eq. (3.6), for an ideal gas

$$\mu_{JT} = \frac{1}{C_p}\left[T\left(\frac{v}{T}\right) - v\right] = 0 . \tag{3.8}$$

An ideal gas would experience no change in temperature in a throttled or Joule–Thomson expansion. Therefore a positive or negative Joule–Thomson coefficient must arise from the departure of real gases from perfect gas behaviour.

Recall that enthalpy is defined as

$$H = U + pV . \tag{3.9}$$

Substituting this in Eq. (3.1) results in

$$\mu_{\mathrm{JT}} = -\frac{1}{C_p}\left\{\left(\frac{\partial u}{\partial p}\right)_T + \left[\frac{\partial (pv)}{\partial p}\right]_T\right\} . \tag{3.10}$$

The first term in Eq. (3.10) is indicative of the departure of a real gas from perfect gas behaviour in terms of Joule's law. This states that the internal energy of a gas is a function only of temperature. For real gases the first term is always negative and contributes to a decrease in temperature on expansion. This arises because the molecules of the gas become increasingly separated as expansion proceeds and their microscopic potential energy is increased. No external work or heat transfer occurs, so the increase in potential energy must be offset by a decrease in the microscopic kinetic energy of the gas molecules and is manifest by a decrease in temperature.

The second term in Eq. (3.10) is indicative of the departure of a real gas from perfect gas behaviour in terms of Boyle's law. This states that the product of pressure and volume of a perfect gas is a function only of temperature.

At low pressures and at temperatures near saturation conditions gases are more compressible than Boyle's law predicts since attractive forces are acting to condense the gas. This means the second term in Eq. (3.10) is negative and contributes to the production of a temperature decrease. At high pressures the molecules are closer together and repulsive forces are brought into action. The gas is therefore less compressible than Boyle's law predicts, and the second term in Eq. (3.10) is positive. This contributes to a temperature increase on expansion.

Increase or decrease in gas temperature on expansion depends upon the relative magnitude of the two terms in Eq. (3.10). Along the inversion curve shown in Fig. 3.13 the two terms are equal in magnitude (and opposite in sign). The second term dominates to the right of the inversion curve and the first term to the left of the inversion curve.

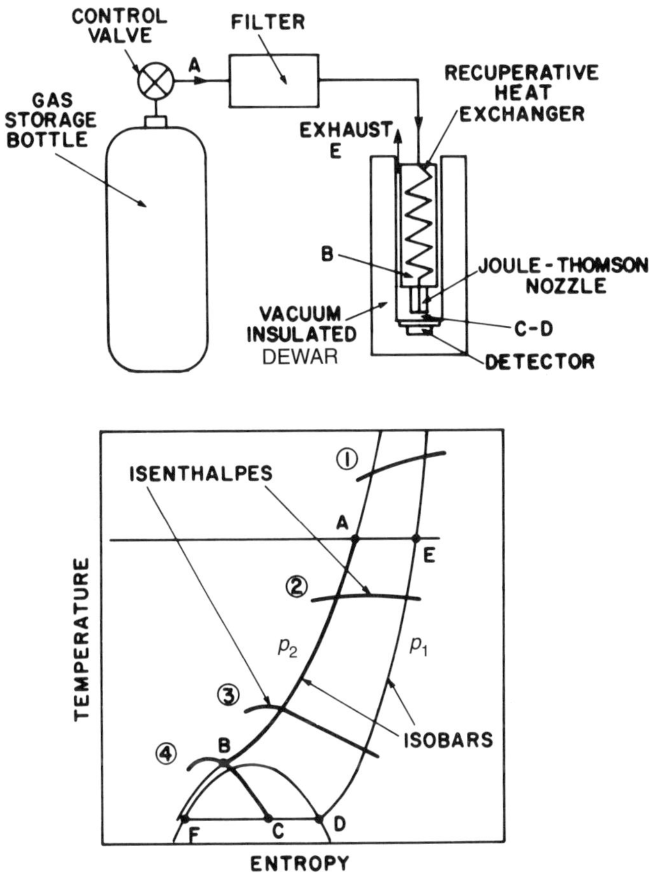

Fig. 3.13 Elements of a stored gas Joule–Thomson expansion cooling system.

3.12 Elements of a stored gas Joule–Thomson cryostat

The essential elements of a stored gas cooling system utilizing Joule–Thomson expansion of the gas are shown in Fig. 3.13. A gas storage bottle is equipped with a control/regulating valve allowing gas at a controlled pressure and room temperature (state A) to pass from the bottle to the gas filter unit and hence to the inlet of a recuperative heat exchanger. In the heat exchanger the gas is cooled from state A to state B by low-pressure gas venting to exhaust and being heated from state D to state E.

The cooled gas then expands isenthalpically through the Joule–Thomson nozzle from state B to state C along the isenthalpe B–C.

Some of the gas is liquefied (to saturated liquid at state F) while the remainder is saturated vapour at state D. The liquid fraction evaporates to state D. In doing so it absorbs heat from the walls of the cold well of the vacuum-insulated dewar encapsulating the heat exchanger and Joule–Thomson nozzle.

The heat absorbed in evaporating the liquid is the refrigeration capacity of the cryostat and cools the detector mounted on a platform secured to the underside of the dewar cold well.

The cold gas at D is vented to atmosphere through the low-pressure channels of the heat exchanger. In passage it is warmed from state D to state E, the heat being drawn from the compressed gas entering and being cooled from A to B.

On Fig. 3.13 various isenthalpes are shown. The isenthalpe labelled 1 is clearly outside the inversion curve for the temperature increases as the fluid expands from p_2 to p_1 The isenthalpe labelled 2 is close to the inversion curve boundary. The fluid temperature remains substantially constant during expansion from p_2 to p_1.

The isenthalpes labelled 3 and 4 are clearly within the inversion envelope for the fluid temperatures decrease on expansion from p_2 to p_1. Note that both these isenthalpes show that expansion from a higher pressure than p_2 (pressure to the left of the p_2 isobar) would, first of all, produce an increase in temperature until the inversion curve was reached (at the apex of the isenthalpe) and a decrease in temperature was obtained.

Gas is stored at pressures well in excess of the optimum pressure for Joule–Thomson expansion, the apex of the isenthalpe labelled 4 on Fig. 3.13. It is expanded from the bottle pressure to the supply pressure p_2 through the control/regulation valve. This initial expansion is itself a Joule–Thomson expansion taking place at room temperature. Depending on the gas it may produce a slight increase or decrease in the temperature of the gas. However the pressure tube connecting the storage bottle to the heat exchanger, through the gas filter, is not thermally insulated and so, therefore, there is ample opportunity for heat transfer with the environment to return the gas to room temperature by the time it arrives at the heat exchanger.

The gas filter used may be a ceramic, metal, or fibre micropore filter or a molecular sieve, arranged in replaceable cartridge form.

3.13 Heat exchangers

The contraflow, recuperative, heat exchanger cooling the compressed gas prior to isenthalpic expansion is a critical component of the system.

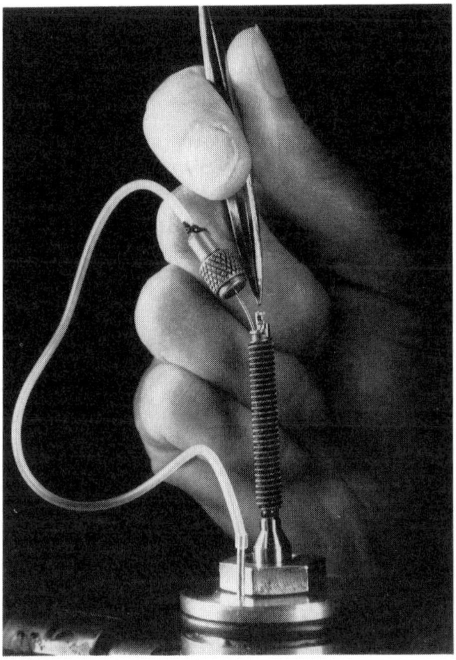

Fig. 3.14 Miniature Joule–Thomson expander with coil-tube Giaque–Hampson heat exchanger (courtesy Hymatic Engineering Co.).

For miniature Joule–Thomson cryocoolers the heat exchanger shown in Fig. 3.14 is widely used. This type of heat exchanger is called a Giaque–Hampson, coil-tube, transverse-flow, recuperative heat exchanger. A length of fine-bore circular tube, usually cupronickel with copper fins soldered to the outside and with a smooth internal bore, is wound in the form of a close-coiled helix around a cylindrical mandrel.

The mandrel may be a thin-wall metal (stainless steel) tube or low-conductivity plastic rod. Typically the outside diameter of the heat exchanger tube is 0.5–1.0 mm (0.02–0.04 in) and the inside diameter 0.3–0.7 mm (0.012–0.028 in) with fins of copper ribbon having a section 0.25–0.5 mm (0.010–0.020 in) by 0.05–0.15 mm (0.002–0.004 in) soldered to the tube in a helical coil of 20–40 turns per cm (50–100 turns per inch). The finned tube is wound on the mandrel with 18–30 turns so the outside diameter of the fins is 5–10 mm (0.2–0.4 in) and the coil is 5–7.5 cm (2–3 in) long.

A cotton or plastic thread spacer is usually wound in the helical interstices between the tube coils. This partially fills the space between the tubes causing the fluid to flow over a longer path with a reduced area thus increasing the velocity and consequently the heat transferred from the fins.

The coiled tubes may be contained in an outer sheath but more usually the assembly is simply inserted in a close-fitting well provided in the glass or metal dewar. The internal walls of the well in the dewar vessel provide the sheath around the coiled-tube heat exchanger. European infrared systems have well bores (outside diameter of heat exchanger coil) of 7.22 mm (0.283 in) while the well bore of US dewars is 5.18 mm (0.204 in).

The compressed gas flows through the coiled tube. The convective heat transfer coefficient between the compressed gas and the tube is very high because of the high density and high velocity of the compressed gas.

The gas expands through the Joule–Thomson nozzle at the cold end of the heat exchanger located at the bottom of the dewar well and the point most distant from the end of the coil tube where the compressed gas enters the tube. In the expansion process taking place at the cold end it is likely some of the gas will be liquefied. The cold liquid droplets impact the closed end of the dewar well on the other side of which is mounted the detector to be cooled. Very high rates of heat transfer are achieved by the liquid gas boiling at the end of the cold well so that highly effective cooling of the detector is achieved.

The evaporated liquid and the other cold expanded low-pressure gas then flows back along the well, over the tube coil, towards the warm end of the heat exchanger. It flows transversely over the coiled tube and the convective heat transfer coefficient is low because of the low density of the gas. This is partially compensated by the increased area for heat transfer of the external fins.

The heat exchanger tube coil fits snugly in a cylindrical well contained within the dewar causing the return gas to remain in close contact with the finned tube as it flows toward the warm end. Heat is transferred from the compressed gas flowing inside the tube, cooling it as it moves toward the cold end of the heat exchanger where the expansion will take place. The heat is transferred to the cold gas flowing transversely over the tube, heating it to ambient temperature as it passes from the cold end to the hot end of the heat exchanger coil.

3.14 Joule–Thompson expansion nozzles

Isenthalpic or Joule–Thomson expansion of the high-pressure cold gas is accomplished at the cold end of the heat exchanger using a variety of devices which may be broadly classified into two groups:

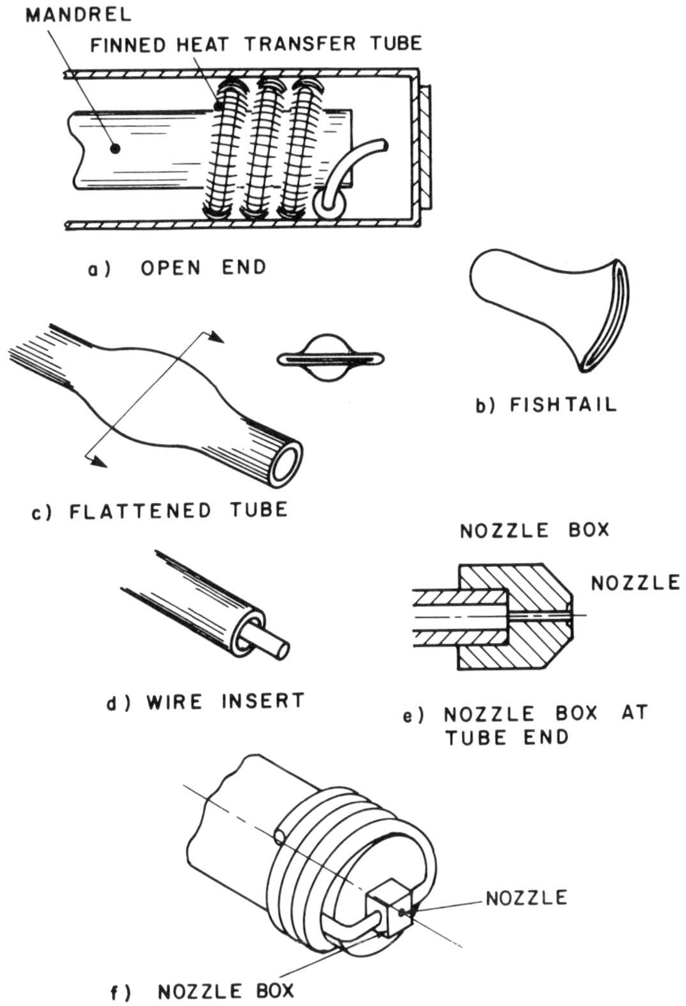

Fig. 3.15 Fixed-orifice Joule–Thomson expansion nozzles.

(1) fixed-orifice nozzles;

(2) demand flow nozzles.

3.14.1 Fixed-orifice expansion nozzles

Some of the fixed-orifice expansion nozzles used in the past or presently in use are shown diagrammatically in Fig. 3.15. The simplest system, shown on Fig. 3.15(a) has no special provision at the cold end of the tube apart from, perhaps, bending it to direct the jet of cold gas issuing forth to impinge on the bottom of the dewar well where the cooled detector is located. The tube end is simply cut square or oblique to the axis. The frictional pressure drop along the heat exchanger tube is the Joule–Thomson expansion. It is a system that works well if the heat exchanger tube is long with a fine bore.

Usually a shorter tube of larger bore is preferred. This requires a restriction at the cold end to limit the mass flow to an acceptable rate and provide the high-friction device, the Joule–Thomson nozzle, to dissipate the pressure energy of the fluid.

One simple way (Fig. 3.15(b)) is to flatten the ends of the tube to a fishtail shape leaving a narrow rectangular slit orifice through which the pressure gas expands. A modification of this is to flatten the tube by rolling near the cold end, Fig. 3.15(c).

Yet another variation is shown in Fig. 3.15(d) where a wire is inserted in the cold end of the tube. The wire diameter is somewhat smaller than the bore of the tube leaving an annular clearance between the wire and the tube walls for flow of the gas. Very precise and reproducible control of the fluid flow can be attained with this technique. The wire diameter and the length of insert are the flow control variables. The mass rate of flow of the gas depends on the cube of the clearance in the annulus, and is a linear function of the depth of insertion of the wire in the tube.

In Fig. 3.15(e) nozzle box is soldered on the cold end of the tube with a fine-bore expansion nozzle drilled axially through the box. Great precision in flow can be obtained in this way.

A variation of the nozzle box method is shown in Fig. 3.1(f). The nozzle box is soldered to the pressure tube a short (5–10 mm) distance from the end of the tube. The end is flattened to close it and anchored by soldering it to the mandrel. The expansion nozzle is formed by drilling a fine-bore hole through the nozzle box and wall of the pressure tube.

Fixed-orifice expansion nozzles have the advantage of simplicity and are easy and cheap to make. However, there are significant difficulties in their use making them unsuitable in many applications.

The biggest problem is that contaminants in the compressed gas will liquefy and solidify during the expansion thus blocking the very fine orifice of the expansion nozzle. Water is the most common contaminant. Others may include carbon dioxide, hydrocarbon vapours from the compressor lubricant, solvents used in plastics, varnishes, and the like. Complete elimination of all the contaminants is not usually practicable unless extraordinary measures are taken to clean the gas downstream of the compressor before it is stored in compressed gas bottles.

Another disadvantage of fixed-orifice expansion nozzles is that the gas flow is precisely opposite to that desired. To illustrate, consider first the continuity equation for flow in a duct:

$$m = \rho Ac \tag{3.11}$$

where m is the rate of mass flow of the fluid, σ is the fluid density, A is the cross sectional area of the duct, and, c is the velocity of the fluid.

The equation of state for the gas states that

$$p/\rho = zRT \tag{3.12}$$

where p is the fluid pressure, z is the compressibility factor, R is the characteristic gas constant of the fluid, and, T is the temperature of the fluid.

It is clear from these equations that:

(1) the density of the fluid increases as the temperature decreases;

(2) as the density increases the mass rate of flow through the orifice or duct increases proportionately.

At start-up, when the entire device is at ambient temperature, the density is relatively low so the mass flow will also be relatively low. We would much prefer the mass flow to be high at the start to produce maximum cooling and a rapid cool-down to the operating condition.

After start-up the cold end of the heat exchanger becomes cold, the density of the gas at the lower temperature increases, and so the mass flow is high. This is contrary to the ideal condition where the mass flow decreases as the unit cools to its steady operating condition. The refrigeration rate necessary to maintain the unit at its operating condition is likely to be much less than that required to cool the unit initially from ambient temperature.

The flow rate in a practical, fixed-orifice, Joule–Thompson cryostat is therefore precisely opposite to the theoretical ideal. If the unit is designed to have a high rate of flow, and hence a quick cool-down at start-up, the rate of flow thereafter will be excessive and wasteful, producing much more refrigeration than is necessary to maintain steady operating conditions. This is an important consideration for miniature Joule–Thomson systems supplied from a storage bottle of limited volume capacity. As the pressure in the bottle decreases the gas density will decrease with consequent reduction in the rate of mass flow.

Another factor of great importance is the ambient atmospheric temperature, the temperature at which gas enters the cryostat. The available refrigeration generated per unit mass of gas is progressively reduced as the ambient temperature increases. The fixed cryostat must be adjusted so the mass flow is sufficient to generate the minimum necessary refrigeration at the maximum specified ambient temperature. At all lower ambient temperatures the mass flow of gas will be excessive and wasteful, generating more refrigeration than is necessary.

The accumulation of condensed contaminants at the expansion orifice will partially block the orifice and so tend to limit the flow. However, this is not a precise and effective control mechanism. In practice, the condensate accumulates to block the orifice entirely thus stopping the fluid flow and the refrigeration process. Eventually the expansion nozzle increases in temperature sufficiently for the plug of condensed contaminate to melt or become dislodged enough for the gas pressure to blow it out of the orifice. This process of intermittent and unregulated operation of the cooler results in unacceptable temperature variations of the cooled detector assembly.

In summary, a fixed-orifice cryostat is simple and can be made at relatively low cost but the flow characteristics are not ideal so that excess gas is consumed. The need for clean contaminant-free air is paramount and any savings in the nozzle costs are frequently eliminated entirely by the increased costs for compressor, bottle, and filter systems.

3.14.2 Demand flow nozzles

The difficulties and deficiencies of fixed-orifice expansion nozzles led, in the early 1960s, to the invention of rapid cool-down, demand flow or throttling expansion nozzles. In these systems the expansion orifice is open to the maximum extent at ambient temperature. A high rate of gas flow is, therefore, experienced at start-up with a consequent rapid cool-down.

As the cold-end temperature decreases the expansion orifice is progressively closed thus reducing the flow rate. At operating temperature, usually when some of the expanded gas is liquid, the orifice may be entirely closed with perhaps provision for a secondary fixed orifice to allow a small maintenance flow.

Increase of the cold-tip temperature will cause the control mechanism to open the expansion orifice. Fluid is allowed to pass in sufficient quantity to restore the cold tip to the operating temperature, with liquefied gas probably present, when the orifice will again be closed. The gas consumption of demand flow cryostats is then very much less, some cases only 10 per cent of the gas flow for the equivalent fixed orifice Joule–Thompson cryostat.

Accumulation of the condensed contaminant at the expansion orifice can be readily accommodated by demand flow expansion nozzles. As the contaminant accumulates to limit and perhaps stop the flow the control mechanism simply increases the orifice opening to maintain the gas flow.

This feature relaxes very considerably the need for the gas to be absolutely contaminant free. Savings on the size, weight, and cost of the gas clean-up equipment offsets the extra cost and complexity of the demand flow nozzle.

A further significant advantage of the demand flow nozzle is that it is virtually independent of the supply pressure of the gas. The control parameter for the demand flow nozzle is the temperature of the cold tip, not the gas supply pressure. Therefore, as the supply pressure decreases, the amount of refrigeration produced per unit mass decreases, and so the fraction of time the demand flow nozzle is closed decreases, but the temperature at the cold tip remains constant throughout.

The reduced gas flow of the demand flow cryostat has the significant advantage that the noise level, sometimes called the *microphonic characteristic*, of the demand flow expansion nozzle is significantly less than the fixed-orifice nozzle.

Similarly, vibrations induced by the gas flow could be expected to reduce simply because the gas flow is less. However, the 'white noise' spectrum of an expanding gas jet has many frequency and harmonic components. There is always the possibility that one of these components may correspond to the natural frequency or a harmonic of some part of the structure and so excite a substantial mechanical vibration.

Tests results given by Buller (1971), reproduced in Fig. 3.16, show comparative flow rates of nitrogen required to produce 0.2 W refrigeration in test assemblies with commercially available fixed-orifice and demand flow cryostats. Buller compared the theoretical performance of optimized demand flow and fixed-orifice expansion nozzles with nitrogen and found, for the particular case he considered, a 50 per cent improvement in the operating time for a system with a limited-size storage bottle. He commented, however, that the actual improvement experienced in practice is more of the order of a 500 per cent increase in the operating time.

The reason for the difference is that the fixed-orifice cryostat cannot operate continuously at the optimized state Buller assumed for the theoretical comparison. As we saw above the orifice must be adjusted for the highest ambient temperatures so at all other ambient temperatures the flow will be excessive. Moreover, at start-up, the mass flow increases as the cold-tip temperature decreases so, at the operating condition, the mass flow will be excessive and wasteful.

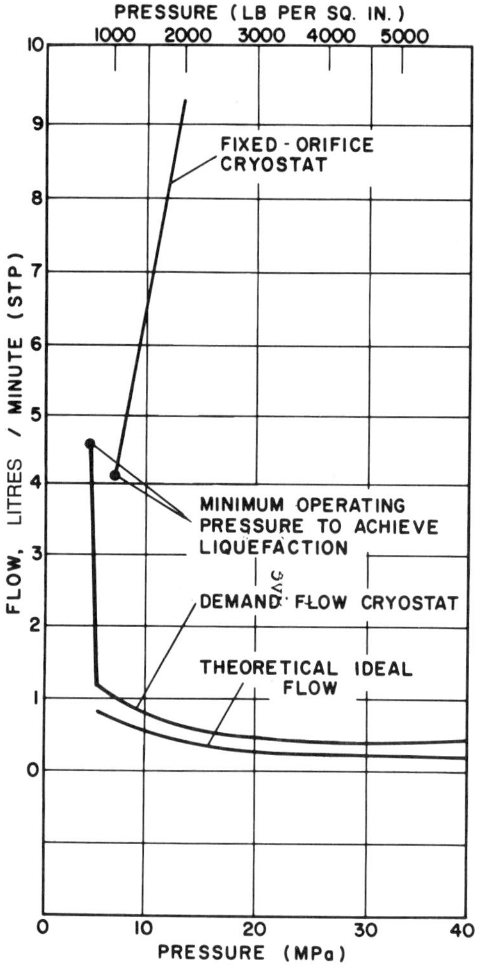

Fig. 3.16 Comparative nitrogen flow rates for 0.2 W refrigeration with fixed-orifice and demand flow expansion nozzles (after Buller 1971).

Furthermore, the fixed orifice cannot be adjusted to clear frozen contaminant and so its operation is limited to relatively low inlet pressures with correspondingly higher flows.

All these factors, the need to cater for high ambient temperature, the need for adequate mass flow at start-up, and the limitation to relatively low pressures because of contamination, combine to result in excessive and wasteful mass flows with correspondingly high velocities and consequently lower heat exchanger efficiencies.

As Buller says: 'The inability of the non-throttling cryostat to compensate for changes in the heat load or for a gradual clogging tendency requires a highly excessive initial flow setting to obtain reliable long-term operation. Thus, it is the throttling cryostat's ability to respond to changing conditions, and the inability of the non-throttling crysotat to operate near optimum, which accounts for the relatively good performance of the throttling cryostat.'

Several ingenious mechanisms and procedures to achieve a demand flow expansion nozzle are described in the literature, particularly in patents. Three systems currently in use are described below.

Santa Barbara Research Center/Air Products The system currently used in the Santa Barbara Research Center (SBRC) and Air Products (APD) demand flow cryostats is shown in Fig. 3.17. The high-pressure tube is coiled on the mandrel with the cold end arranged very similar to the fixed-orifice, nozzle box, type shown in Fig. 3.15(f). The gas flow through the orifice is controlled by a valve needle which enters the orifice. The needle is attached to a small metal bellows contained within the cold end of the mandrel. As the temperature of the cold end decreases the pressure of gas in the bellows decreases allowing the extended bellows to retract causing the needle to move into the orifice and restrict the gas flow.

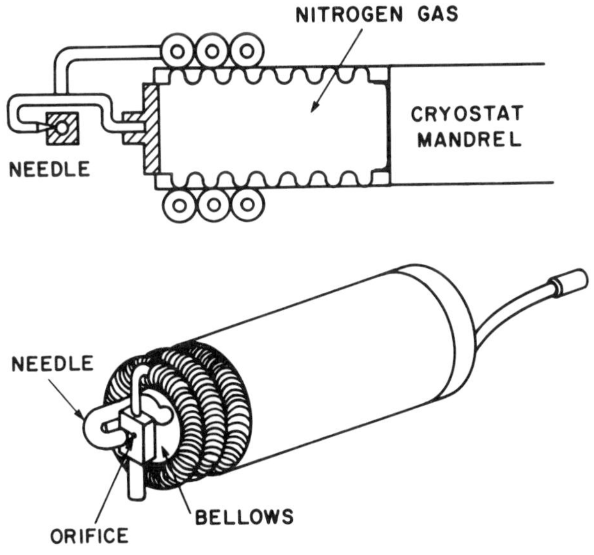

Fig. 3.17 SBRC/APD demand flow crysotat (after Buller 1971).

The bellows is usually charged with refrigerant gas at supercritical pressure. When it is controlling the gas flow through the orifice a fraction of the bellows charge is liquid so the pressure in the bellows is a function only of the temperature at the cold end, not the initial charge pressure of the bellows.

The bellows is thermally attached to the mandrel very close to the cold end so that it functions at a temperature some 30 °C above the normal boiling point of the refrigeration at the low expanded gas pressure. It is fortunate the vapour pressure, the pressure of fluid above a boiling liquid, increases with temperature. This provides the possibility for some of the bellows charge to be liquid (at the higher bellows pressure) even though the temperature of the bellows is above the temperature of the liquefied expanded gas in the cryostat.

In some systems intended for use with nitrogen refrigerant gas (normal boiling point of nitrogen is 77 K) the bellows have been filled with a mixture of nitrogen and argon gas (normal boiling point of argon is 88 K). This accentuates the reduction in bellows charge pressure as the temperature drops allowing the stretched bellows to retract thus causing the needle to enter the orifice and restrict the gas flow.

The bellows mechanism does not respond directly to the presence of liquid refrigerant gas at the detector but to a temperature somewhat above the liquid temperature. It can, therefore, act to increase the flow before all the liquid is lost at the detector. Moreover, operation of the control mechanism is virtually insensitive to gravity and the cryostat may be used in any orientation.

The bellows are very small for they must be accommodated within the diameter of the mandrel. Thus a cryostat having an outside diameter of 5 mm (0.02 in) with a tube outer diameter of 0.5 mm (0.02 in) with fins 0.25 mm (0.01 in) long would have a mandrel diameter no more than 3 mm (0.118 in) diameter.

The bellows are made by electroless nickel plating an aluminum form subsequently dissolved away by immersion in sodium hydroxide. It is said that quality control and reproducibility in production of the bellows has been a persistent problem so that individual adjustment of the needle setting is necessary. Some users have reported the bellows after use sometimes appear to become porous and lose the charge pressure sufficiently to render the unit inoperable. There is little published information about the operating life and failure modes of the SBRC/APD demand flow cryostat.

Hymatic self-regulating minicooler Details of the Hymatic self-regulating minicooler were disclosed by Stephens (1970). A schematic diagram of

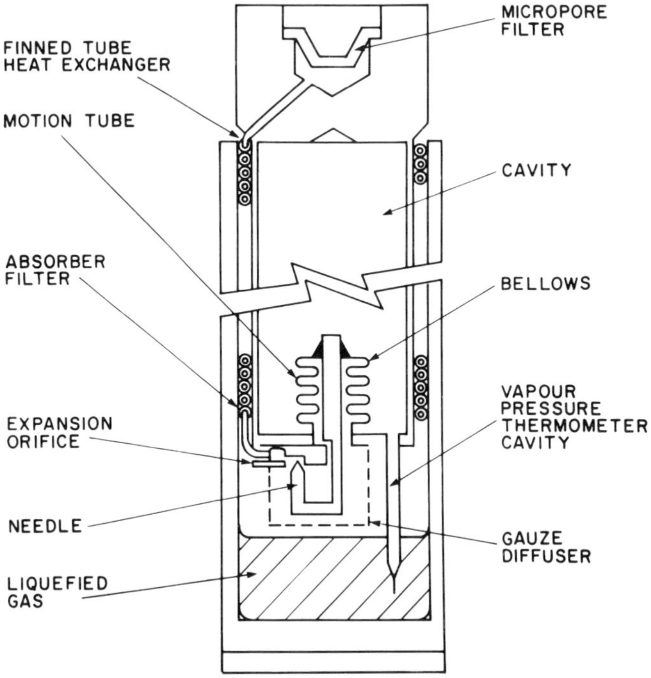

Fig. 3.18 Hymatic self-regulating minicooler (after Stephens 1970).

the unit is shown in Fig. 3.18. The flow of gas through the expansion orifice is regulated, as before, by the motion of a needle moving in or out of the orifice. The flow area of the orifice is automatically varied by a vapour pressure thermometer device so the cold produced just balances the heat loads on the device to maintain liquid gas temperature in the bottom of the dewar well.

The orifice needle is secured to a small bellows coaxial with and intruding into a chamber or cavity contained within the mandrel and charged with the refrigerant gas. A small fill tube or thermometer bulb connected to the chamber extends beyond the mandrel and is sealed shut so as to contain the gas charge in the chamber.

The end of the thermometer bulb is immersed in the liquefied refrigerant gas downstream of the expansion orifice and located on the back face of the detector. Gas contained in the thermometer bulb will be liquefied so the charge pressure in the cavity will always be the vapour pressure corresponding to the temperature of the liquefied gas pool in the minicooler.

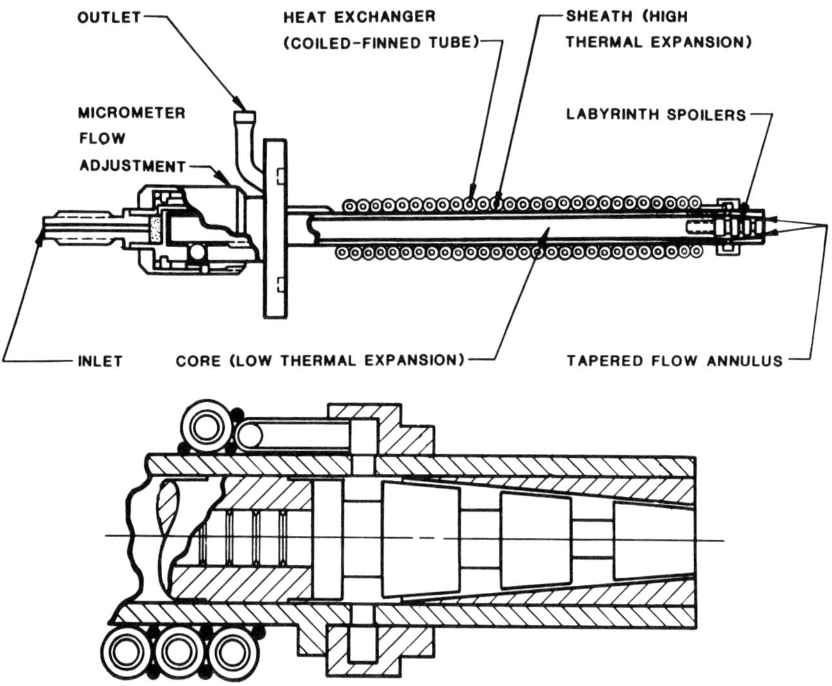

Fig. 3.19 General Pneumatics variable-area temperature-sensitive expansion nozzle (after Hedegard *et al.* 1986).

As the charge pressure in the chamber is decreased the force squeezing the needle bellows is relaxed allowing the bellows to extend into the chamber carrying the needle into the orifice and restricting the gas flow.

Comparing the two systems it can be seen that the bellows in the Hymatic system is pressure balanced whereas the SBRC/APD bellows is internally pressurized. The absence of a significant pressure difference across the Hymatic bellows suggests it is unlikely to suffer from the problem of leaky bellows noted above and, therefore, its usable life is likely to be significantly longer.

A disadvantage of the Hymatic system may be that, because it depends on the presence of liquid refrigerant gas around the thermometer bulb, it is likely to be sensitive to attitude orientation and the high acceleration effects which occur in aerial combat, missile firing, and the like. No information on this aspect of its use appears to have been published.

Changes in the ambient pressure, such as a reduction in atmospheric pressure with increase in elevation, impinge on the operation of Joule–Thomson systems in that the low post-expansion vapour pressure is correspondingly reduced and hence the temperature of the liquid refrigerant gas.

Such changes in the ambient pressure are automatically compensated by the vapour pressure thermometer feature of the Hymatic cooler. They are not compensated for in the bellows system of the SBRC/APD systems discussed previously. The reduction in the post-expansion pressure will tend to allow the bellows to stretch further thus moving the needle out of the orifice so increasing the gas flow.

General Pneumatics variable-area temperature-sensitive Joule–Thompson expansion nozzle A different type of demand flow Joule–Thomson expansion nozzle was developed by the General Pneumatics Corp., Scottsdale, Arizona. This is illustrated in Fig. 3.19. The finned heat exchanger tubing containing the high-pressure gas is helically wound on a mandrel in the usual way. The mandrel is in two parts:

(1) a core of material having the lowest possible coefficient of thermal expansion (such as Invar, a 34 per cent nickel–iron alloy, or G-10 epoxy–fibreglass plastic);

(2) a sheath around the core of a material having the highest possible coefficient of thermal expansion (aluminum).

The cold end of the core carries a truncated cone of Invar having an included angle of 15 to 20°. This fits into a corresponding tapered section at the cold end of the sheath.

The cold end of the pressure tube is secured to the sheath at a point close to the cold end where there is a hole in the sheath allowing the tube to pass through it. The cooled gas leaving the pressure tube then expands isenthalpically in the tapered annulus formed by the tapered, male, end of the core in the corresponding tapered, female, end of the sheath.

At start-up there is a relatively large annular gap allowing maximum gas flow. As the temperature of the cold end decreases the sheath, with a high coefficient of thermal expansion, retracts relative to the core so reducing the annular gap and, thereby, restricting the gas flow.

In some versions the sheath and the core are rigidly connected at the warm end. In other versions a micrometer screw adjustment at the warm end permits relative axial movement between the core and sheath. This is useful to adjust the flow rate to changing conditions and may be done either at the start of operation during initial set-up, or when the unit is operating.

The assembly can be contained in an insulated outer sheath with a liquid reservoir at the cold end and can be simply inserted in the cavity of a vacuum-insulated dewar detector vessel in the customary fashion.

A series of circumferential grooves, 'labyrinth spoilers', cut in the tapered, male, end of the core, were provided to accentuate frictional dissipation of the fluid flow. Hedegard *et al.* (1987) found the labyrinth spoilers had no effect on the characteristics of the flow but served a useful function of catchment reservoirs for condensed contaminants resulting from the expansion process.

The General Pneumatics expansion nozzle is presently in limited production and has considerable potential for widespread application. It is a relatively robust device lending itself to mass production at low cost. Moreover, it has demonstrated the capability to accommodate relatively highly contaminated gases for long periods without mishap. This is because the expansion orifice is annular in form and unlikely to become simultaneously blocked. The stringent need for clean refrigerant gases to avoid contaminant blockage is therefore relaxed with consequent savings on the gas filter and clean-up systems.

3.15 Working fluids in Linde–Hampson cryocoolers

Virtually any gas can be expanded isenthalpically in a Linde–Hampson system to produce a refrigeration effect. Some gases, such as neon, hydrogen, and helium, have inversion temperatures less than ambient and so must be precooled before Joule–Thomson expansion.

While miniature hydrogen liquefiers, with an associated nitrogen precooler, have been made to cool detectors to 20 K (McInroy 1967) they are not common. However, there appears to be an increasing demand for the use of two-stage, nitrogen/hydrogen liquefiers. Even three-stage nitrogen/hydrogen/helium cooling systems are in course of development.

Most gases have maximum inversion temperatures greater than the ambient and so can be expanded to produce refrigeration without pre-cooling. Different results are obtained with different gases.

Table 3.3, reproduced from Barron (1984), shows the performance of different fluids operating in an ideal Linde–Hampson system between maximum and minimum pressure limits of 0.1–20 MPa (15–3000 lb per sq in) and an initial temperature of 300 K (80 °F). It was assumed the compression process was isothermal and the heat exchanger effectiveness was 100 per cent.

Table 3.3. Performance of the Linde–Hampson system using different fluids. $p_1 = 101.3$ kPa (14.7 psia); $p_2 = 20.265$ MPa (200 atm); $T_1 = T_2 = 300$ K (80 °F); heat-exchanger effectiveness = 100 per cent; compressor overall efficiency = 100 per cent (after Barron 1985).

	Normal boiling point		Liquid yield $y = \dot{m}_f/\dot{m}$	Work per unit mass compressed		Work per unit mass liquefied		Figure of merit FOM = $\dot{w}_i/\dot{w}$
Fluid	K	°R		kJ/kg	Btu/lb$_m$	kJ/kg	Btu/lb$_m$	
N_2	77.36	139.3	0.0708	472.5	203.2	6673	2869	0.1151
Air	78.8	142	0.0808	454.1	195.2	5621	2416	0.1313
CO	81.6	146.9	0.0871	468.9	201.6	5381	2313	0 1428
A	87.28	157.1	0.1183	325.3	139.8	2750	1182	0.1741
O_2	90.18	162.3	0.1065	405.0	174.1	3804	1636	0.1671
CH_4	111.7	201.1	0.1977	782.4	336.4	3957	1701	0.2758
C_2H_6	184.5	332.1	0.5257	320.9	138.0	611	262	0.5882
C_3H_8	231.1	416.0	0.6769	159.0	68.4	235.0	101.0	0.5976
NH_3	239.8	431.6	0.8079	363.1	156.1	449.4	193.2	0.7991

The table is arranged in order of increasing value of the normal boiling point temperature. The liquid yield is the mass fraction of gas involved in the Joule–Thomson expansion converted to liquid as a result of the expansion. The work per unit mass compressed and the work per unit mass liquefied are self explanatory. The figure of merit is defined as the ratio of the theoretical minimum work requirements (to liquefy unit mass of gas, say) divided by the actual work requirement of the system. The theoretical minimum work requirement is deduced from a hypothetical cycle, not achievable in practice, wherein the gas is compressed isothermally to such a high pressure that, at the end of isentropic expansion, the gas is all saturated liquid. The 'actual work' requirement of the system was calculated from an idealized cycle in which the compression process was assumed to be isothermal and with perfect heat exchange processes. A practical system with friction losses, limited heat transfer, radiation and conduction losses, and non-ideal compressors would have a much higher actual work requirement, several times that of the ideal Linde–Hampson cycle.

It is clear from Table 3.3 that in general as the normal boiling point temperature increases the liquid yield and figure of merit increase and the work per unit mass compressed or liquefied decreases.

3.15.1 Argon versus nitrogen

Argon is a notable exception to the general trend shown in Table 3.3. With its normal boiling point of 87 K argon has both a higher liquid yield and figure of merit and a lower work per unit mass compressed or liquefied than its closest neighbours in the table, carbon monoxide at 82 K and oxygen at 90 K.

Compared with nitrogen (boiling at 77 K) the performance of argon (boiling at 87 K) as a refrigerant gas appears to be very superior indeed. The liquid yield of argon is 0.12 compared with 0.07 for nitrogen, and the figure of merit is 0.17 compared with 0.11. The work per unit mass of gas compressed and liquefied are 325 kJ/kg and 2750 kJ/kg respectively for argon and 472 kJ/kg and 6673 kJ/kg for nitrogen.

Buller (1971) confirmed the advantage of argon compared with nitrogen with publication of the figure reproduced in Fig. 3.20 (converted to metric units). The figure shows the refrigeration generated, in watt-minutes per litre of gas flow, as a function of the gas inlet pressure. Buller assumed the ideal cycle with a perfect heat exchanger. Gas temperatures at inlet and exhaust were assumed to be 300 K The inlet gas volume was quoted in litres at standard temperature and pressure (273 K and 1 atm).

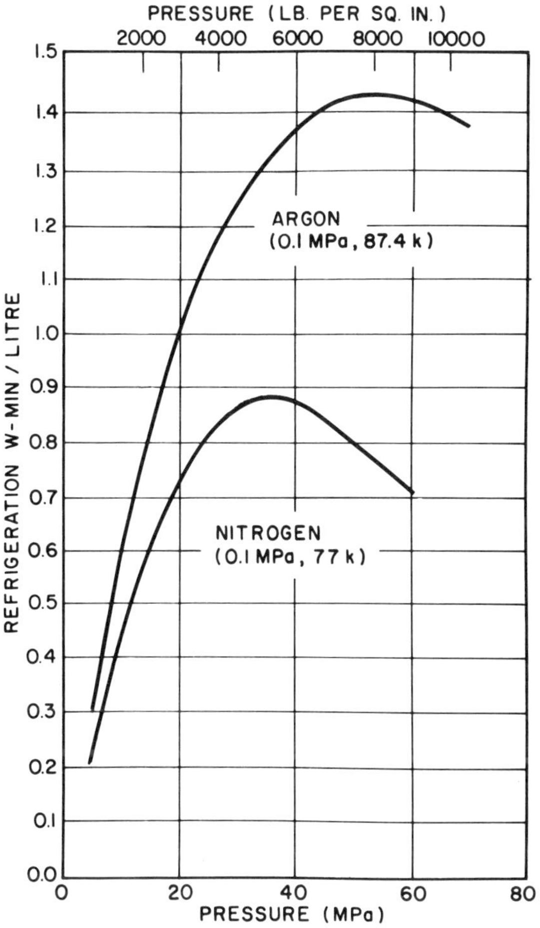

Fig. 3.20 Refrigeration as a function of inlet pressure for nitrogen and argon (after Buller 1971).

Earlier Giest and Lashmet (1960) drew a comparison of argon with nitrogen when used as a refrigerant gas in Linde–Hampson systems. They considered a Linde–Hampson cycle operating between pressure limits of 10 and 0.1 MPa (1500 and 15 lb per sq in).

With heat exchangers designed to have a warm-end temperature difference of 5.5 °C (10 °F) between the inlet and outlet gas temperatures they observed that the nitrogen heat exchanger needed to be almost 50 per cent larger than the corresponding argon exchanger. They concluded also that, with a specific heat exchanger and a given volumetric flow rate, the refrigeration available from the argon would be greater than from the nitrogen.

Both nitrogen (atomic number 14, molecular weight 28) and argon (atomic number 39.9, molecular weight 39.9) are present in atmospheric air and are produced commercially in air separation plants. Nitrogen is the major constituent of air, 78 per cent by volume, 75 per cent by weight. Argon is also present in air but in a concentration of only 0.93 per cent by volume and 1.25 per cent by weight. Both gases are readily available from the commercial suppliers at high levels of purity. Argon is more expensive than nitrogen but fortunately the price difference is not in the same ratio as their relative abundance. A recent quotation for small quantities of the two gases, industrial grade purity better than 99.5 per cent, indicated the argon at $92 per 9.15 m^3 compressed gas bottle was two-and-a-half times the price of nitrogen at $27 per 6.21 m^3 compressed gas bottle.

The principal use of argon at present is the inert shroud gas of the welding process known as argon arc welding. Nitrogen is widely used as a low-cost inert gas for a wide variety of purposes.

3.15.2 Air

Air can be used as the working fluid or refrigerant gas for miniature Linde–Hampson cryocoolers and has the compelling advantage that it is the most abundant gas on earth and is readily available everywhere at no cost (except of course for underwater (naval submarines) or space applications).

Air is principally composed of nitrogen (78 per cent by volume, 75 per cent by weight) and oxygen (21 per cent by volume, 23 per cent by weight) with many other constituents, mainly water vapour, argon, carbon dioxide, and a wide variety of other trace elements.

Normal atmospheric air contains a considerable quantity of water vapour. The actual amount depends on both the temperature and on the humidity. 'Absolute humidity' is defined as the mass of water vapour contained in unit mass of dry air. When the air contains all the water vapour it can carry it is said to be saturated. The actual pressure of the water vapour in saturated air depends on the temperature and can be determined by reference to steam tables. For example, at 20 °C the vapour pressure of the water in saturated air is 2.2 kPa and the mass of water vapour per unit mass of dry air, the absolute humidity, is 0.014 kg H_2O/kg dry air.

The relative humidity of air is the ratio of the mass of water actually contained in a given mass or volume of dry air to the mass of water contained in saturated air. 'Bone-dry' air (no water vapour present) has a relative humidity of zero, saturated air has a relative humidity of 100 per cent. In desert (Arizona in summer) or high, dry, cold regions (Calgary in winter) the relative humidity may be as low as 10–15 per cent. In warm coastal regions (New Orleans or Houston) or in many areas following rain the humidity can be very high, 90 per cent relative humidity at temperatures in the range 20–25 °C.

When normal atmospheric air is compressed in a compressor to a pressure of, say, 100 times the pressure at which it was inhaled a considerable amount of liquid water will be precipitated. Assume, for example, that air is inhaled to the compressor at a temperature of 20 °C with a relative humidity of 50 per cent. With a relative humidity of 50 per cent the actual mass of water vapour associated with 1 kg of dry air is 0.007 kg, half the amount that would be associated with saturated air.

Now let the air be compressed to 100 times the original atmospheric pressure and cooled in the intercooler and aftercooler heat exchangers back to the original temperature 20 °C. The volume of the air will be approximately 100 times less than the original volume because the pressure has increased 100 times.

In this process, the water vapour has not been compressed 100 times, for the vapour pressure depends only on the temperature and at the end the temperature is 20 °C, the same as at the start of the compression. However, the volume now available for the vapour is only 1/100th of the original volume. Had the air been saturated at the start, 99 per cent of the water originally present would have precipitated as liquid.

Since the air was only 50 per cent saturated at the start it would not have become saturated until the volume was reduced to half the original value. Further reduction in volume would result in the precipitation of water. At the end when the final volume was only 1/100th of the original volume the air would be saturated but would contain only 1/50th of the original water; the other 49/50ths would have precipitated as liquid water.

Liquid water can be easily removed from the air. Most of the liquid will separate automatically from the air by gravity and can be retained by a water trap. Fine droplets of liquid water suspended in the air can be removed by a relatively simple filter.

This disposes of most of the water (98 per cent) entering the compressor in the inlet air. Nevertheless the compressed, saturated air leaving the compressor still contains water in the form of vapour. There is no simple way to remove it and so it passes into the heat exchanger where the air is cooled to a very low temperature before expanding in the Joule–Thomson nozzle to the original atmospheric pressure.

In the cooling process the temperature of the air decreases to a very low value. In cooling, most of the water vapour precipitates from the air as solid, frozen, water on the inner walls of the heat exchanger high-pressure tube coiled around the mandrel. The velocity of the air through the tube may be so high that some of the precipitate remains suspended in the air.

A similar conversion of vapour to liquid or solid precipitates will occur with the carbon dioxide and other vapours, principally organic, that may be present in the atmospheric air.

At the end of the cooling process the majority of contaminants in the original air will have precipitated in the heat exchanger. Nevertheless, some will still remain in the vapour state despite the very low temperature and vapour pressure of the cooled stream.

The further reduction in temperature experienced during the Joule–Thomson expansion in the nozzle will result in further precipitation of the water vapour and other contaminants still remaining in the gas. These condensed contaminants, coupled with the suspended solids carried along by the air following their earlier precipitation in the heat exchanger, aggregate in the fine orifice of the expansion nozzle. Fixed-orifice cryostats are particularly vulnerable to blockage by contamination. This problem can be largely resolved by the use of demand flow cryostats which automatically open to permit the cold gas to flow even when there is appreciable contaminant build-up at the orifice. The area for flow of a fully open orifice in a demand flow cryostat can be many times the area of the fixed orifice and so an appreciable contaminant build-up can be tolerated.

3.15.3 Mixed fluids

In a recent publication Little (1987) has discussed the expansion of gas mixtures, nitrogen and hydrocarbons with halon (presumably to suppress the fire hazard), with very great improvement in the efficiency allowing substantial reduction in the operating pressure (from 13 MPa (2000 lb per sq in)) to a third of this value. In addition the increased refrigeration capacity resulted in astonishing reductions in the cool-down time. Little included the figure reproduced in Fig. 3.21 showing the 'cool-down of a typical refrigerator ... both with the use of nitrogen and the mixed gas, in each case with a supply pressure of 1500 lb per sq in (10 MPa)'. With the gas mixture the device cools below 100 K in a little over a minute. With small changes in the channel depths of the refrigerator, a cool-down time of tens of seconds is obtainable. The volumetric flow at this pressure is about 0.6 litre per min (gas at standard temperature and pressure).

These results confirm the improvement in performance to be gained through the use of mixed gases first observed by Alfeev *et al.* (1973).

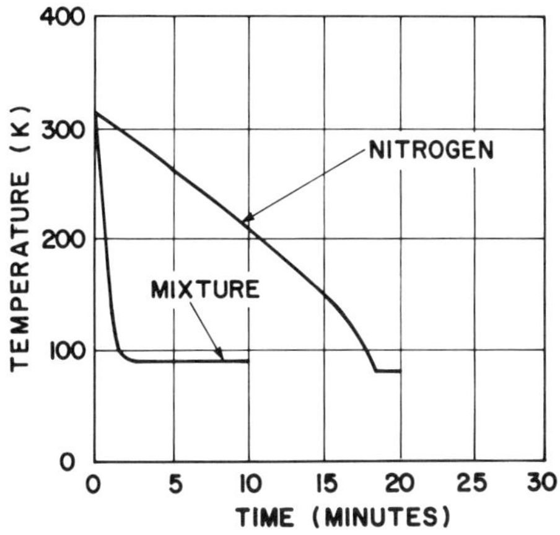

Fig. 3.21 Typical cool-down curve for refrigerator with nitrogen and special gas mixture (after Little 1987).

The improvement is of great importance. It radically enhances the prospects for potential application of Joule–Thomson cryocoolers particularly in the closed-cycle version. Difficulties in the engineering development of reliable low-capacity compressors would be substantially relaxed if it were possible to operate at one-third the pressure necessary with nitrogen, argon, or air. A one-day symposium on mixed gas flows (Jones 1990) brought together an assembly of manufacturers and users of Joule–Thompson systems and represents a cameo snapshot of the state of development of this field.

Little and Paugh (1991) reported on the use of mixed fluids for a fast cool-down Joule–Thomson microminiature cryocooler for infrared focal plane arrays. Maytal *et al.* (1991) reported on an extensive theoretical thermodynamic analysis of mixed fluid Joule–Thomson cryocoolers. This milestone theoretical paper was later supplemented by experimental work reported by Maytal (1993).

3.16 Rapid cool-down single-shot miniature cryocoolers

Smart projectile munitions require miniature cryocoolers with special characteristics that make Joule–Thomson minicoolers the only feasible cooling system. They must have an extremely rapid cool-down time: about 2.5 seconds from the moment the gun is fired, or the missile launched, to the state where the thermal sight in the nose of the projectile is actively seeking a target. During this cool-down phase they must be able to withstand extremely high accelerations, up to 50,000 *g*, and a very broad spectrum of vibration. The required operating life is limited to a maximum of 60 seconds. They need to have an unlimited storage life with no maintenance required, and be instantly ready for use without preparation, at all times, under the widest possible range of operating conditions, such as ambient temperatures ranging from −40 °C to 50 °C. Finally they should operate with great reliability, and be low in cost. Miniature, bottle storage Joule–Thomson cryocoolers seem to meet all these requirements.

Figure 3.22 shows a rapid cool-down, single-use, sealed-for-life Joule–Thomson cryocooler, the MAC 219 series, in production at the Hymatic Engineering Company. This unit uses argon refrigerant gas with an initial storage bottle pressure of 400 atm.

Air or nitrogen can also be used as the refrigerant gas but the cool-down time is generally four times longer than with argon.

Hart (1984) reviewed the development of fast cool-down cryocoolers at the Hymatic Engineering Company. The minimum temperature achieved is that of the boiling liquid downstream of the expansion nozzle. This temperature depends on the pressure of the refrigerant vapour over the boiling liquid and the pressure, in turn, depends on the ambient pressure and the pressure drop resulting from the heat exchangers restricting flow on the low-pressure side; the larger the gas flow, the higher the pressure drop, and the higher the minimum liquid temperature.

For a given cold end diameter (the detector mount diameter) it is found that a lower temperature can be obtained faster with cone-shape heat exchanger coils rather than the customary parallel cylinder coil. Lower temperatures and cool-down times are attained as the included cone angle is increased, but at the expense of a larger mass flow of gas due to increase in the heat exchanger mass.

Requirements for fast cool-down and long run times are mutually opposed. Fast cool-down requires low heat exchanger thermal capacity, attained with a short length of heat exchanger tubing and hence a low heat exchanger efficiency. Extended run times require a high heat exchanger efficiency, hence a long heat exchanger tube with high heat capacity to achieve the minimum gas consumption. As in most applications, therefore, the minicooler design must be a compromise of conflicting requirements including the selection of a fixed-orifice or demand flow Joule–Thomson nozzle, the greater efficiency of the one against the simplicity and lower cost of the other.

The encapsulation associated with the minicooler has an important effect on the cool-down time. A reduction in the encapsulation thermal mass reduces the refrigerant gas flow required for a given cool-down

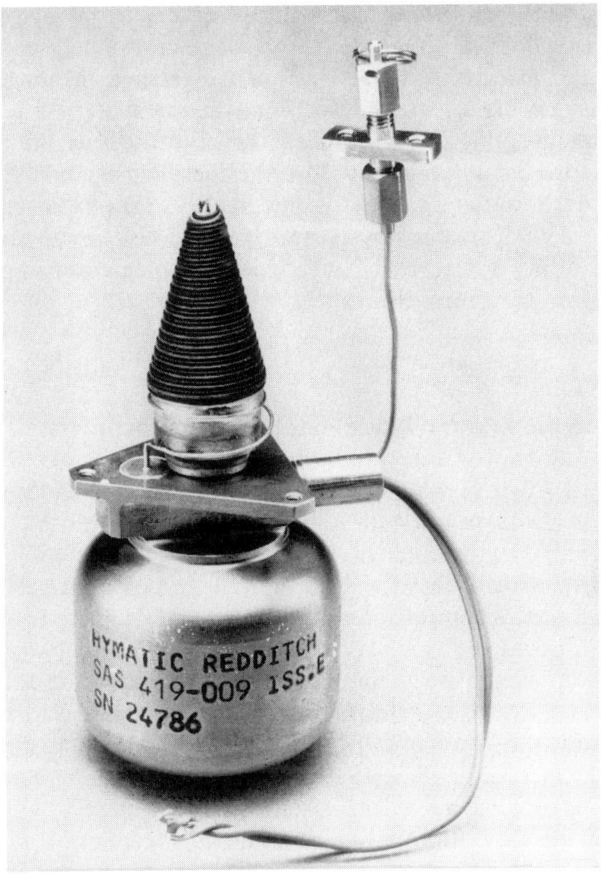

Fig. 3.22 Rapid cool-down, single-use, sealed-for-life cryocooler MAC 219 series (courtesy Hymatic Engineering Co.).

time and this, in turn, leads to a shorter heat exchanger giving a shorter encapsulation and an overall reduction in system size.

An alternative form of miniature autonomous rapid cool-down cryocooler at an early stage of development at the General Pneumatics Corporation, Scottsdale, Arizona, is shown in Fig. 3.23. The entire system is contained in a spherical reservoir made in two parts and circumferential seal welded. The encapsulation, detector, and optics assembly are contained in a cylindrical insert screwed into the reservoir and electron beam welded shut.

The cryocooler heat exchanger and Joule–Thomson expansion nozzle is located at the lower end of the cylindrical insert and separated from the compressed gas in the reservoir by a diaphragm across the lower face of the cylindrical insert.

An inertia valve is contained within the tapered mandrel of the heat exchanger coil. The inertia valve consists essentially of a heavy mass support on a spring. Operation of the cryocooler is initiated when a needle carried in the lower face of the inertia mass punctures the diaphragm to allow the compressed gas in the reservoir to enter the heat exchanger coil. This process is anticipated to occur at the instant of launch or firing when very high acceleration loads are imposed on the system. It is presumed the sphere will be located at the nose of the missile or shell.

3.17 Microminiature Linde–Hampson cryocoolers

In the late 1970s a new type of Linde–Hampson cryocooler was invented by Professor William Little of Stanford University, Palo Alto, California. Little perceived a requirement for cryocoolers having cooling capacities very much less than the miniature units described above. He recognized that it was not technically feasible to scale existing units to the level required. Instead he and his coworkers applied a photolithographic process to fabricate, on a flat glass plate, the very fine heat exchanger channels and expansion nozzles required. They were successful in this development and later founded the company MMR Technologies Inc. to seek commercial applications for their work.

The following is an extract from a recent paper by Little (1987) describing the new Joule–Thomson expander combined with cryogenic electronic systems (reprinted with permission).

3.18 Microminiature Joule–Thomson refrigerators

Joule–Thomson refrigerators produce refrigeration by expanding high-pressure gas, which has been precooled in a counter-current heat exchanger, through a porous plug or capillary. The outgoing cool gas passes up the same heat exchanger and precools the incoming gas. These coolers differ from other mechanical refrigerators in that they have no moving parts in the cold end of the refrigerator. This is important in many applications where low noise is a primary consideration.

Miniature Joule–Thomson refrigerators with heat exchangers wound of coiled, finned tubing have found extensive use for cooling infrared detectors. These are bulky and expensive to fabricate. Recently, microminiature refrigerators have been developed which are fabricated in a totally different manner. These are made by photolithographically etching fine gas channels in thin glass laminates to form the counter-flow heat exchanger and expansion capillary. These laminates are then fused together to give a solid, planar refrigerator. The fabrication process, which is clearly related to the fabrication methods of the semiconducting industry, allows one to design refrigerators of arbitrary shape and size. Widescale use is now being made of these refrigerators for laboratory and instrumentation applications.

The combination of this technology with Cerdip packaging allows one to envisage a totally integrated device package in which a microminiature refrigerator provides the refrigeration of the logic device which is mounted on the cold end of the refrigerator inside a vacuum, or foam-insulated ceramic package. Such a device would offer the engineer the performance advantages of a cryogenic chip with the convenience of conventional packaging.

Joule–Thomson refrigerators have normally been considered to be relatively inefficient compared to Stirling cycle or G–M refrigerators. This is due to the irreversible expansion of gas in the Joule–Thomson process. However, as shown first by Alfeev *et al.* (1972), large improvements in efficiency in J–T coolers can be made by the use of mixtures of gases, rather than with a single gas such as nitrogen (see Fig. 3.23). We have confirmed this and shown that suitably formulated mixtures, containing halon, hydrocarbons, and nitrogen, under certain operating conditions, can have refrigeration efficiencies of the order of 50 per cent to 80 per cent of Carnot in the 80 K range and, in addition, are non-flammable and safe to use. This much greater efficiency simplifies the design of the compressor needed to supply the high-pressure gas. With nitrogen, pressures of the order of 2000 psi are required, but for the mixtures, we have demonstrated operation at 750 psi. Recent calculations of the phase diagram of more complex mixtures suggests that operation at 450 psi would be possible with a suitably redesigned refrigerator.

Compressors in this relatively low pressure range can be expected to be able to operate continuously for periods of 10,000 to 20,000 hours without maintenance. This, together with the capability of miniaturizing these refrigerators makes the overall system advantage of these refrigerators substantial. For the first time it is now both feasible and practical to consider integrating the cryogenics with the electronic chip in a single ambient temperature package. This offers the electronics design engineer an attractive option for accessing the superspeed performance and other advantages of cryogenic operation without the pain of working with an immersible circuit board.

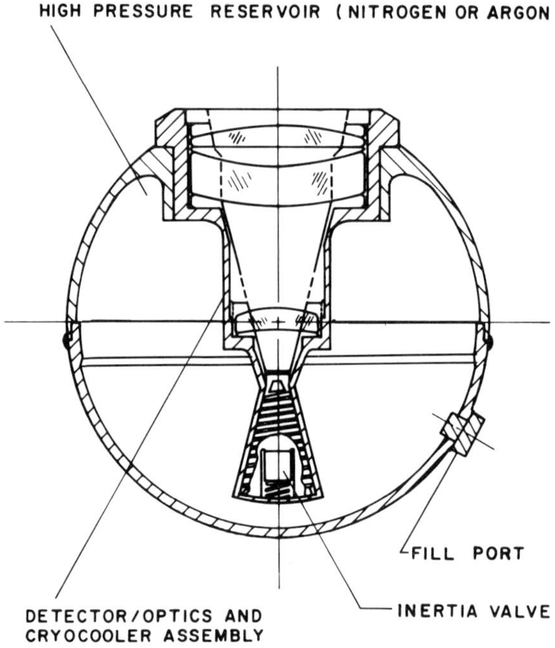

Fig. 3.23 Miniature autonomous rapid cool-down cryocooler (courtesy General Pneumatic Corp.).

3.19 Closure

In this chapter we have briefly summarized some aspects of the fundamental technology relating to open-cycle cooling systems. These include a variety of techniques with the common thread that, in all cases, they operate as open-cycle systems wherein the working fluid is vented and lost by atmospheric dispersion. In those cases where the helium is retained in large plastic bags for subsequent reliquefaction it could be considered a closed-cycle operation.

Our survey embraced stored cryogen systems contained in insulated dewars, presently the most widely used form of low-capacity cryogenic cooling. Liquid cryogens, principally nitrogen and helium, are favoured for many applications in a normal gravity environment, while solid cryogen or supercritical storage is used mainly for space applications.

Stored gas systems have the compelling advantage over stored cryogen systems that the gas can be stored for an indefinite period without loss.

Moreover, when required, the cooling effect resulting from expansion of the gas can be produced very quickly and can be specifically targeted, where it is needed, in a precise and carefully regulated amount. Contemporary developments of guided projectiles with rudimentary, heat-seeking, infrared guidance systems will likely increase the number of gas storage systems in use. The recent developments of planar expanders and the great improvements in performance resulting from the use of mixed gases is most appropriate. It comes at the time when interest in cold electronics, both semiconducting and superconducting, is rapidly increasing.

References

Alfeev, V. N., Brodyansky, V. M., Yagodin, V. M., Nikolsky, V. A., and Ivatsov, A. V. (1972). British Patent Application No. 1336892, Nov. 14.

Barron, R. (1984). *Cryogenic Systems*, 2nd edn. Monographs on Cryogenics 3, Oxford University Press.

Buller, S. (1971). A Miniature Self Regulating Rapid Cooling Joule–Thomson Cryostat. *Advances in Cryogenic Engineering* (ed. K. Timmerhaus), **16**, 205–13.

CRYODATA (1993). CRYODATA computer programs (CCPAK and HEPAK) available from CRYODATA, P. O. Box 558, Niwot, Colo. 80544.

Donabedian, M. (1978). Cooling Systems. Chapter 15 of *The Infrared Handbook* (ed. L. W. Wolfe and G. J. Zissis), Infrared Information and Analysis Center, Environmental Research Institute of Michigan, Ann Arbor, Mich.

Geist, J. M. and Lashmet, P. K. (1960). Miniature Joule–Thomson Refrigeration Systems. *Advances in Cryogenic Engineering* (ed. K. Timmerhaus), **5**, Plenum, New York. (See also: Geist, J. M. and Lashmet, P. K. (1961). Compact Joule–Thomson Refrigeration Systems. *Advances in Cryogenic Engineering* (ed. K. Timmerhaus), **6**.

Hart, R. R. (1984). *Minicooler Systems for Fast Cooldown Applications*. Technical Memorandum TM 1341, Hymatic Engineering Co. Ltd., Redditch, Worcester, UK.

Hedegard, K., Walker, G., and Zylstra, S. (1987). Temperature Sensitive, Variable Area Joule–Thomson Expansion Nozzle. *Proceedings of the Fourth International Cryocooler Conference*, Easton, Md, David Taylor Naval Ship R. and D. Center, Annapolis, Md.

Jones, J. (ed.) (1990). *Mixed Gas Flows: A Symposium*. Jet Propulsion Laboratory, California Institute of Technology, Pasadena, Ca.

Little, W. A. (1984). Microminiature Refrigeration. *Review of Scientific Instruments*, **55**, 661–80.

Little, W. A. (1987). Integrated Cryogenic Chip Carrier for High Speed CMOS. *Low Temperature Electronics Symposium*, (ed. F. Gaenslen), Electrochemical Society, New Jersey, Conference at Honolulu, Hawaii, Oct.

Little, W. A. and Paugh, R. L. (1991). Development of a Fast Cooldown Joule–Thomson Microminiature Refrigerator and Vacuum Package for Operation of Infrared Focal Plane Arrays at 70 K. *Proceedings of the Sixth*

International Cryocooler Conference (ed. G. Green and M. Knox), David Taylor Naval Ship R. and D. Center, Annapolis, Md.

Longsworth, R. (1978). Concepts for Cooling Small Superconducting Devices Using Closed Cycle Regenerative Refrigerators. In *Applications of Closed Cycle Cryocoolers to Small Superconducting Devices* (ed. J. E. Zimmerman and T. M. Flynn), NBS Special Publication 508, US Government Printing Office/NTIS.

Maytal, B. (1993). Experimental Verification of Joule–Thomson Cryocooler Cooldown Periods Similarity. *Proceedings of the Seventh International Cryocooler Conference* (ed. J. Ludwigsen), Kirtland, AFB Santa Fe, N.M.

Maytal, B., van Sciver, S. W., and McMahon, P. D. (1991). Thermodynamic Analysis of Mixed Fluid Joule–Thomson Cryocoolers. *Proceedings of the Sixth International Cryocooler Conference* (ed. G. Green and M. Knox), David Taylor Naval Ship R. and D. Center, Annapolis, Md.

McInroy, J. (1967). Miniature Joule–Thomson Coolers. *Proceedings of the XII International Congress of Ref./Prog. in Ref. Sc. Technol.* **1**, pp. 59–68, Madrid, (IIR, Paris).

Radebaugh, R. (1993). Course notes for the annual course (Refrigerators for Cryogenic Sensors and Cold Electronics). Division of Continuing Education, University of California, Los Angeles 90024–2883.

Sherman, A. (1983). NASA Needs and Trends in Cryogenic Cooling. In *Refrigeration for Cryogenic Sensors* (ed. M. Gasser), NASA Conference Publication 2287, US Government Printing Office/NTIS.

Stephens, S. (1970). A Self Regulating miniature Joule–Thomson Refrigerator. *Applications in Cryogenic Technology* (ed. R. Vance), **3**, 119–203.

Timmerhaus, K. D. and Flynn, T. M. (1989). *Cryogenic Process Engineering*. International Cryogenics Monograph Series, Plenum, New York.

Walker, G. (1983). *Cryocoolers*. 2 vols., Plenum, New York.

4

MECHANICAL CRYOCOOLERS. PART 1: REGENERATIVE SYSTEMS

4.1 Introduction

The open-cycle expendable fluid coolers considered in the previous chapter are satisfactory for many applications and have many advantages. However, there are other needs and requirements for cryogenic cooling which cannot easily or conveniently be satisfied by stored cryogen or gas systems.

These include a variety of long-term, continuous or intermittent, routine, cryogenic refrigeration applications at different temperature levels, principally 4 to 10 K, 20 K, and 80 K with capacity requirements ranging from fractions of a watt at 4 K or below to a kilowatt or more of cooling at 80 K.

The replenishment of liquid cryogen baths or stored gas bottles is tedious at best, often difficult, and sometimes impossible to accomplish. There are many areas of the world where the liquid cryogens are not readily available and, just as important, the skills necessary to transfer them are also not readily available. Sometimes the location of the point of use is inaccessible.

The need has therefore become evident for a range of electrically driven or thermally activated mechanical refrigerators capable of achieving and sustaining cryogenic temperature levels. Such machines should be safe, reliable, and easy to use, requiring no special skills beyond the ability to close a switch or press a button. They should function without maintenance for long periods, be unobtrusive with little or no impact on the environment, and be small, compact, without mechanical vibration or electromechanical emissions. They should have a low initial and operating cost, and have the ability to operate at any required temperature, refrigerating capacity, and orientation without regard to ambient conditions or imposed vibration or environmental conditions. Unhappily such ideal cryogenic refrigerators exist only in Camelot.

The vapour-compression refrigeration units used in domestic and commercial refrigerators go a long way to fulfilling the above specifications. However, unfortunately, they cannot achieve cryogenic temperatures in the uncomplicated single-stage, arrangements so familiar for use at near ambient temperature, such as in domestic refrigerators and air conditioning systems. There are simply no suitable refrigerant fluids capable of operating over the temperature range from ambient to cryogenic. Multiple-stage machines using different fluids, such as Freon, propane, and methane, at progressively lower temperatures with the refrigeration of one stage precooling that of another are possible but are complicated and have not reached the same stage of advanced commercial development of their higher-temperature brethren.

Over the past 40 years much effort and funding has been expended to develop cryogenic mechanical refrigerators, cryocoolers, that satisfy, in one way or another, some of the requirements enumerated above. The purpose of this and the next chapter is to briefly review some of the cryogenic refrigerators which have passed into commercial service or which continue as promising research development projects.

4.2 Mechanical cryocoolers: classification

Newcomers to the field of mechanical cryocoolers find there is a bewildering array of different thermodynamic cycles, system arrangements, and mechanical configurations for cryogenic refrigeration. To bring some order and understanding of this diversity it is useful to divide the different mechanical arrangements into two groups:

(1) the *regenerative systems* which incorporate a regenerative heat exchanger and experience pulsatile reversing flow. These include Stirling, Vuilleumier, Gifford–McMahon, Solvay, Postle, and pulse-tube refrigerators;

(2) the *non-regenerative systems* which have recuperative but not regenerative heat exchangers and which experience steady continuous flow. These include Claude, Linde–Hampson, Joule–Thomson, and Joule–Brayton refrigerators.

The regenerative systems are considered below, the non-regenerative systems in the next chapter.

4.3 Regenerative heat exchangers

Regenerative heat exchangers comprise a porous matrix usually of cylindrical or annular form, of finely divided material, such as fine-diameter metal wires, woven mesh, powders, or balls. They have a single set of flow channels through the matrix, basically the interstices between the packing elements. The hot fluids and the cold fluids must therefore flow sequentially, usually in opposite directions, called *contraflow*. This has the significant incidental advantage that contaminants carried in by one of the fluids and caught in the matrix tends to be blown out by the other fluid proceeding in the reverse direction. Regenerators in contraflow are therefore, to some extent, self-cleaning.

During the *hot blow* the warmer fluid flows through the matrix with heat transfer from the fluid to the matrix. The fluid is cooled in heating the matrix.

After some time the flows are switched and the direction of flow reversed. The supply of warm fluid to the hot end of the matrix is stopped and a supply of cold fluid to the cold end of the matrix is started. This is the start of the *cold blow* during which fluid passes through the matrix with heat transfer from the matrix to heat the fluid. After a suitable interval the cold flow is discontinued and the flows are again switched and reversed to start the hot blow.

The regenerative matrix may thus be thought of as a thermal flywheel or thermodynamic sponge alternately absorbing and releasing heat to or from the matrix in the same way that water is released and absorbed by squeezing and releasing a sponge.

Now in most forms of regenerative cryocooler there is only one working fluid flowing, not separate flows of hot fluid and cold fluid as implied above. The single fluid flows in pulsatile fashion, first in one direction then in another, to and fro in the regenerative matrix like the tide rising and falling on the seashore. In fact this type of flow is called *tidal regenerative flow*.

The frequency of flow reversals occurs at twice the frequency of operation of the cryocooler (two reversals per cycle). In some machines the speed is so high that none, or very little, of the working fluid ever passes right through the matrix. Instead the heat transfer process is a complicated 'bucket brigade' operation made even more complex by cyclic pressure variations of approximately sinusoidal form.

4.4 Recuperative heat exchangers

All mechanical cryocoolers operate by means of cyclic compression and expansion of the working fluids along with heat absorption at low temperatures (the refrigerating effect) and heat rejection at near ambient temperatures (the cooling process). These heat transfer processes are facilitated by recuperative heat exchangers of one sort or another. All mechanical cryocoolers have recuperative heat exchangers but only some types have regenerative heat exchangers.

A recuperative exchanger is characterized by the provision of individual flow channels, separated by solid walls, for the hot and cold fluids. The flow of these fluids occurs simultaneously and may be continuous, pulsatile, discontinuous, or even reversing flow as the tidal flow described above. Most of the many different types of heat exchangers so widely used in domestic, commercial, and industrial equipment are recuperative (see Walker 1990).

4.5 Regenerative cryocoolers

There are four basic types of mechanical cryocooler which incorporate regenerative heat exchangers. These are generally classified by the thermodynamic cycle on which they operate, specifically:

(1) Stirling;

(2) Vuilleumier;

(3) Gifford–McMahon;

(4) pulse-tube.

There are many variations, mechanical configurations, and alternative systems possible over a wide range of temperatures and refrigerating capacity.

Stirling, Vuilleumier, and pulse-tube cryocoolers are similar in that they have no valves. Instead the flow of working fluid is controlled by internal volume changes. Such regenerative machines are classified, generically, as *Stirling engines* (Walker 1983).

Gifford–McMahon cryocoolers and the similar Solvay and Postle machines do have valves which regulate the flow of working fluid. Such systems are often classified, generically, as *Ericsson engines* (Walker 1983).

4.6 Ideal cycle of operation: the Carnot cycle

The Carnot cycle is the thermodynamically ideal refrigerator operating between given temperature limits, T_C (ambient temperature) and T_R (refrigerating temperature). The cycle and idealized engine is shown in Fig. 4.1. The two columns of diagrams to the left of the figure are the pressure–volume (work) and the temperature–entropy diagrams for the

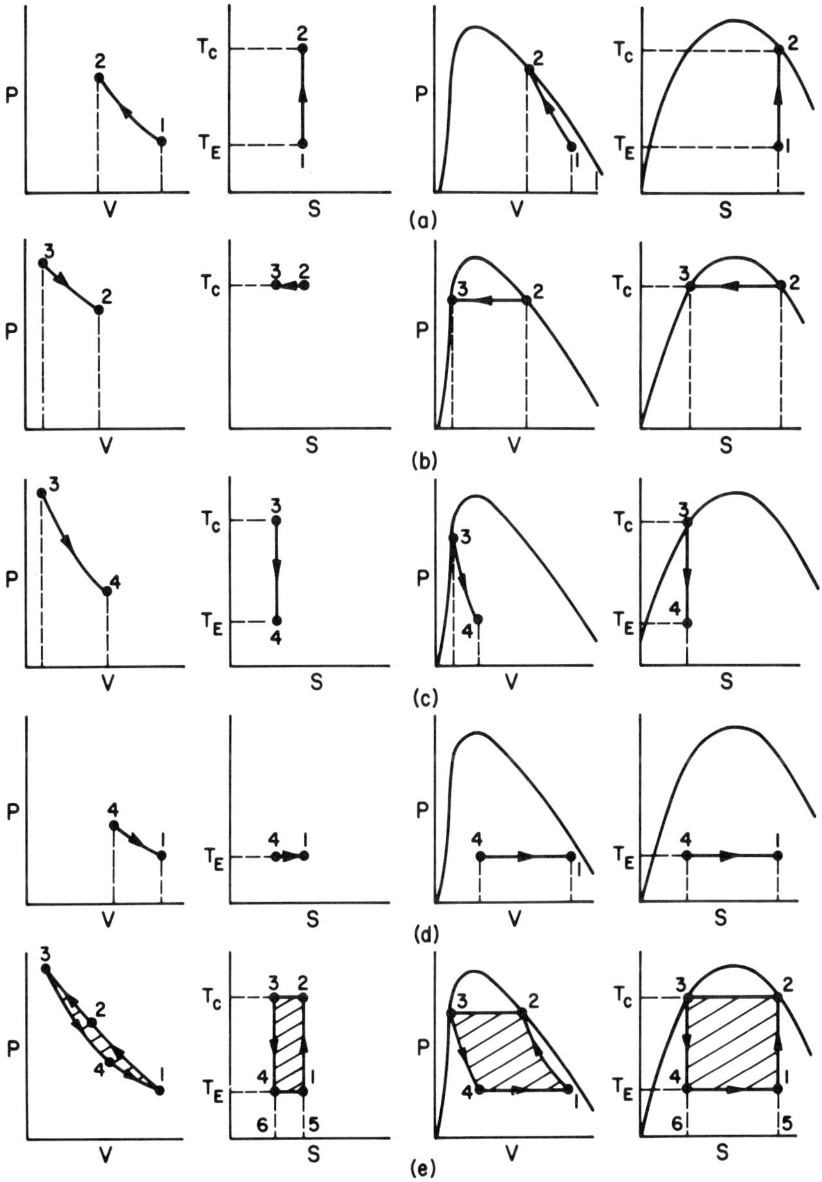

Fig. 4.1 Ideal Carnot cycle.

Carnot cycle with a gaseous working fluid. The two columns of diagrams to the right of the figure are the corresponding diagrams for the Carnot cycle with a vapour working fluid changing phase from saturated vapour to saturated liquid and vice versa.

A conceptual unit capable of operating on this cycle consists of a closed cylinder containing some working fluid and equipped with a piston reciprocating in the cylinder. It would have a cylinder head possessed of magical properties that would allow it to be either perfectly insulating or perfectly conducting as required at different parts of the cycle.

The thermodynamic cycle of the Carnot refrigerator consists of four processes as follows:

1. Starting with the piston at the outer dead point (maximum volume in the cylinder) the piston moves toward the cylinder head compressing the gas in process (1–2). This is shown on the pressure–volume ($P-V$) and temperature–entropy ($T-S$) plane by the line (1–2). It will be recalled that the areas on the $P-V$ diagram represent work transfer and on the $T-S$ diagram represent heat transfer. In process (1–2) the compression is assumed to be adiabatic, that is, the cylinder head is in its perfectly insulating mode and no heat is transferred to or from the working fluid during the process. The area on the $P-V$ plane during this process is the work necessary to accomplish the compression process, W_{1-2}. On the $T-S$ plane the process is a vertical straight line, there is no area enclosed below the line, so there is no heat transfer in the adiabatic process. The pressure and temperature increase along the isentrope 1–2 (constant entropy, adiabatic), shown on both the $P-V$ and $T-S$ planes.

2. Next, for the isothermal compression (2–3) the cylinder head becomes perfectly conductive. Heat is transferred from the working fluid at the constant temperature T_C while the pressure increases from p_1 to p_2. In this process heat is rejected from the system to the surroundings cooling the working fluid. The pressure increases throughout this isothermal compression process. Work is done by the piston on the gas.

3. At point 3 the piston reaches the end of its stroke and reverses direction to begin the expansion stroke. The cylinder head becomes perfectly insulating once more, and the initial part of the expansion process (3–4) is carried out adiabatically. No heat is gained or lost by the working fluid but the pressure and temperature fall throughout the process. Work is done by the gas on the piston.

4. At point 4 the cylinder head once more becomes perfectly conducting and during process (4–1) heat is transferred from the surroundings to the working fluid at temperature T_R. This is the useful refri-

geration process. The temperature of the working fluid remains constant, the pressure falls, and work is done by the gas on the piston.

It can be seen from the above that the ideal Carnot refrigerator has four processes occurring sequentially (a) two adiabatic processes, one a compression and one an expansion process; and (b) two isothermal processes, one a compression (at temperature T_C) and one an expansion (at temperature T_R).

The two adiabatic processes involve the expansion or compression of the same mass of gas between the same temperature limits, T_C and T_R. Therefore the compression work, process (1–2), must be the same magnitude (but opposite sign; input rather than output) as the expansion work of process (3–4).

The work of the isothermal compression process (2–3), equal to the heat transferred from the gas to the surroundings, may be calculated from:

$$W_{1-2} = Q_C = mRT_C \ln (V_3/V_2) \tag{4.1}$$

where W_{1-2} is the work of isothermal compression in process (1–2), Q_C is the heat transferred at temperature T_C from the working fluid in process (1–2), m is the mass of working fluid in the cylinder, R is the universal gas constant, T_C is the temperature at which the compression occurs, and, (V_c/V_b) is the compression ratio, the ratio of the volumes of the cylinder at the beginning and end of the isothermal compression process.

Similarly the work of the isothermal expansion process (4–1), equal to the heat lifted Q_R at temperature T_R, may be calculated from:

$$W_{4-1} = Q_R = mRT_R \ln (V_4/V_1) \, . \tag{4.2}$$

The net work input to the cycle is therefore:

$$W_{in} = W_{1-2} - W_{4-1} = mR(\mathrm{T}_C - T_R)\ln r \tag{4.3}$$

where

$$r = \frac{V_2}{V_3} = \frac{V_1}{V_4} \, .$$

The coefficient of performance, the ratio of heat lifted, Q_R, to work input W_{in}, is therefore:

$$\mathrm{COP} = \frac{Q_R}{W_{in}} = \frac{T_R}{T_C - T_R} . \tag{4.4}$$

The Carnot cycle has the highest possible coefficient of performance between the given temperature limits T_R and T_C. Referring to the $T-S$ diagram in Fig. 4.1 it is easy to understand why this is the best possible cycle. The area under the isothermal process (4–1) is the heat lifted, Q_R. Given that the temperature at which refrigeration occurs is limited to the maximum value T_R any process other than the isothermal process would decrease the heat lifted and increase the work done. In similar fashion, referring to the heat rejection process (2–3) the heat rejected, Q_C, is the area under the isothermal process (2–3) on the $T-S$ plane of Fig. 4.1. Given that coolant (water or air) is available at temperature T_E the minimum area for a compression process on the $T-S$ diagram will be given by the isothermal process. Any other process would involve temperatures greater than T_C. This would increase the area of the diagram of heat rejected, Q_C.

The net cycle input work is the area of the lowermost diagrams bounded by the four processes and is simply the difference between the heat lifted Q_R and the heat rejected Q_C.

The important principle thus established by the Carnot cycle is that, for maximum performance, all the heat transfers to and from the refrigerator must take place at constant temperature and these constant temperatures must be the limiting temperatures T_C and T_R of the cycle. It is easy to see from Fig. 4.1 that the work input to the refrigerator will increase as the refrigeration temperature is decreased. At the same time the amount of heat lifted, Q_R, will decrease as the refrigeration temperature is decreased.

Thus a second important principle is that refrigeration should never be carried out at a temperature lower than is absolutely necessary.

Having described this excellent machine it is unfortunate that we must now recognize that a practical Carnot refrigerator cannot be made. This is because the slopes of the isothermal and isentropic lines for a gaseous working fluid on the $P-V$ plane are much closer than those shown diagrammatically in Fig. 4.1. To achieve reasonable work and heat transfer quantities very high pressures and large volumes would be necessary, involving a huge, heavy machine. Moreover it is not possible to achieve, in practice, compression and expansion processes that are isothermal or adiabatic. The adiabatic process requires the rate of heat transfer to be zero or the process to occur very fast. The isothermal process requires rates of heat transfer that are infinite, or a process that takes place very slowly. Despite the fact that we cannot actually make a Carnot refrigerator it nevertheless provides a useful theoretical yardstick with which to judge the performance of our practical machines.

A Carnot refrigerator with a vapour working fluid that changes phase from saturated liquid to saturated vapour has the same ideal coefficient of performance as with a gaseous working fluid. The pressure–volume and temperature–entropy diagrams for this case are shown on the right of Fig. 4.1. This cycle also has the great advantage that the isothermal and isentropic processes have very different slopes on the $P-V$ plane and, furthermore, can be closely approximated in practice. Nearly all the vapour compression refrigerators widely used for domestic and commercial near-ambient temperature refrigeration operate on a modified Carnot cycle and achieve about 70–80 per cent of ideal Carnot performance. It is a pity they cannot be used for cryocoolers also but there is no suitable working fluid available. This would need to have a vapour pressure slightly above atmospheric at the lowest temperature (to avoid the problem of air leaking into the unit) and a modest vapour pressure at ambient temperature (to avoid the engineering problem of excessively high pressure).

4.7 Stirling cryocoolers

4.7.1 Ideal cycle for a Stirling refrigerator

The ideal Stirling refrigerator is shown in Fig. 4.2. It consists of a cylinder containing two opposed pistons with a regenerative heat exchanger between them. One of the two volumes between the regenerator and the pistons is called the expansion space, and is maintained at low temperature T_E. The other volume is called the compression space, and is maintained at ambient temperature T_C. There is a temperature gradient (T_C-T_E) across the transverse faces of the regenerator and it is assumed there is no thermal conduction in the longitudinal direction. The pistons are assumed to move without friction or leakage of the working fluid enclosed between them.

Let us assume that initially the compression-space piston is at the outer dead point, and the expansion-space piston is at the inner dead point, close to the face of the regenerator. All the working fluid is then in the compression space at ambient temperature, the volume is a maximum, and the pressure and temperature are at the state (1) on the $P-V$ and $T-S$ diagrams, shown in Fig. 4.2. During compression, process (1–2), the compression piston moves towards the inner dead

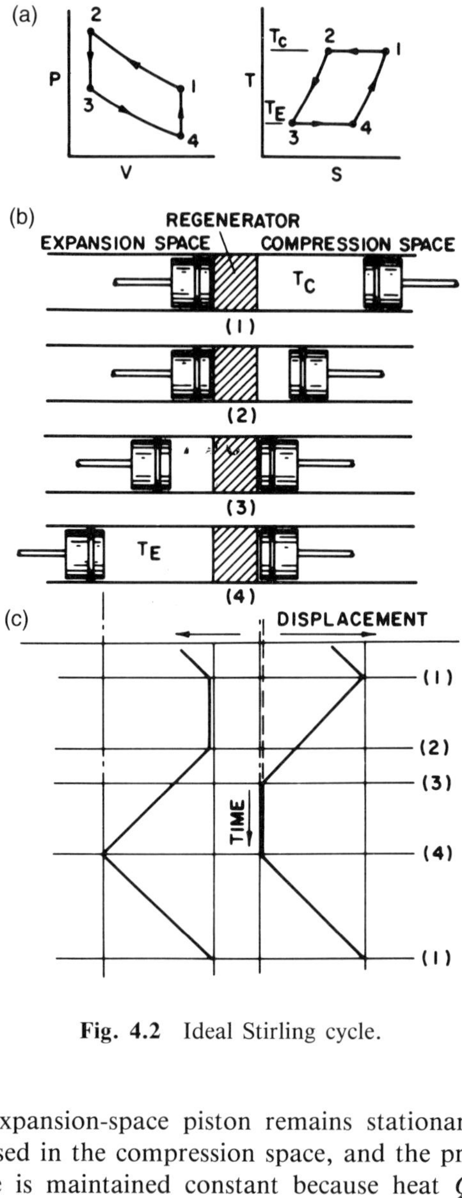

Fig. 4.2 Ideal Stirling cycle.

point and the expansion-space piston remains stationary. The working fluid is compressed in the compression space, and the pressure increases. The temperature is maintained constant because heat Q_C is abstracted from the compression-space cylinder.

In the transfer process (2–3) both pistons move simultaneously, the compression piston towards, and the expansion piston away from, the regenerator. The volume between them remains constant. The working fluid is transferred through the porous metallic matrix of the regenerator to the expansion space. In passing through the regenerator, the working fluid is cooled from T_C to T_E by heat to the matrix, and emerges from the regenerator into the expansion space at temperature T_E. The decrease in temperature passing through the matrix at constant volume causes a decrease in pressure.

In the expansion process (3–4) the expansion piston moves to the outer dead point; the compression piston remains stationary at the inner dead point, adjacent to the regenerator. As the expansion proceeds, the pressure decreases as the volume increases. The temperature remains constant because heat Q_E is abstracted from the cylinder surroundings. This is the useful refrigeration of the cycle.

Finally both pistons move simultaneously to transfer the working fluid (at constant volume) back through the regenerative matrix from the expansion space to the compression space. In passing through the matrix, heat is transferred from the matrix so the working fluid increases in temperature and emerges at T_C into the compression space.

The cycle is therefore composed of four heat-transfer processes:

Process (1–2) Isothermal compression; heat transfer from the working fluid to the external sink at ambient temperature, T_C.

Process (2–3) Constant-volume regenerative cooling; heat transfer from the working fluid to the regenerative matrix.

Process (3–4) Isothermal expansion; heat transfer to the working fluid from an external source at the refrigerating temperature T_E.

Process (4–1) Constant-volume regenerative heating; heat transfer to the working fluid from the regenerative matrix.

The heat transferred in process (2–3) is the same as in process (4–1). The only heat transfer process between the engine and its surroundings are (a) heat supply at T_E and (b) heat rejected at T_C. The fact that the heat is supplied and rejected at constant temperature satisfies the requirement of the second law of thermodynamics for maximum thermodynamic performance. The coefficient of performance of the ideal Stirling cycle is therefore the same as for the Carnot cycle:

$$\mathrm{COP} = \frac{T_E}{T_C - T_E} . \qquad (4.5)$$

The advantage of the Stirling cycle over the Carnot cycle lies in the replacement of two isentropic processes by two constant-volume regenerative processes. This greatly increases the area of the $P–V$ diagram

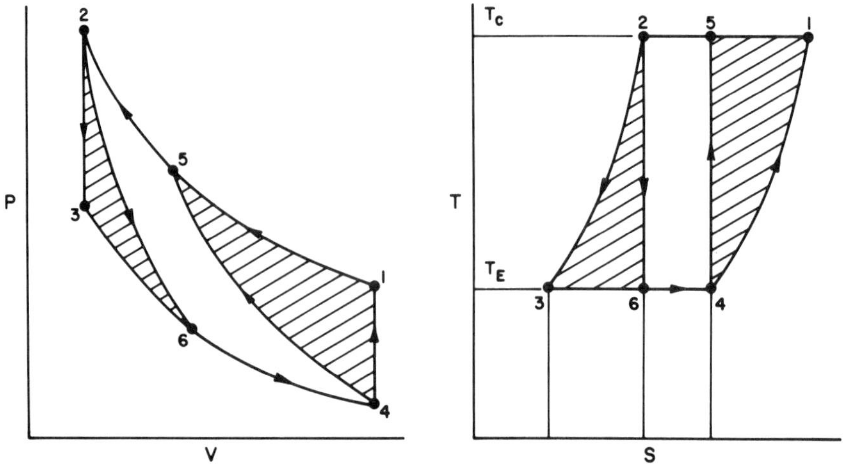

Fig. 4.3 Stirling and Carnot cycles superimposed with common values for the minimum and maximum temperatures, pressures, and volumes. Cross-hatched areas on the *P–V* plane represent the increased work output of the Stirling cycle. Cross-hatched areas on the *T–S* plane represent increased heat transfer of the Stirling cycle.

and the quantities of heat transfer to and from the system so that it is not necessary to resort to very high pressures and swept volumes.

The *P–V* and *T–S* diagrams of the Carnot and Stirling cycles between given limits of pressure, volume, and temperature are compared in Fig. 4.3. The shaded areas, (4–1–5) and (2–3–6), represent the additional work made available by substituting constant-volume processes for isentropic processes. The isothermal processes, (2–5) and (6–4), of the Carnot cycle are extended to processes (1–2) and (3–4), respectively. The quantities of heat transfer for the Stirling cycle are increased by the same proportion as the work. The coefficient of performance (heat extracted per unit of work) is the same for both.

Practical Stirling cryocoolers work on a much more complicated cycle. This has been discussed in detail by Walker (1983, 1989). The best performance that practical Stirling cryocoolers have achieved is a coefficient of performance of 0.45 of the Carnot value.

4.7.2 *The Stirling power system or heat pump*

The Stirling engine is unusual in that the same machinery can operate not only as a *refrigerator* but also as a *power system* or *heat pump*.

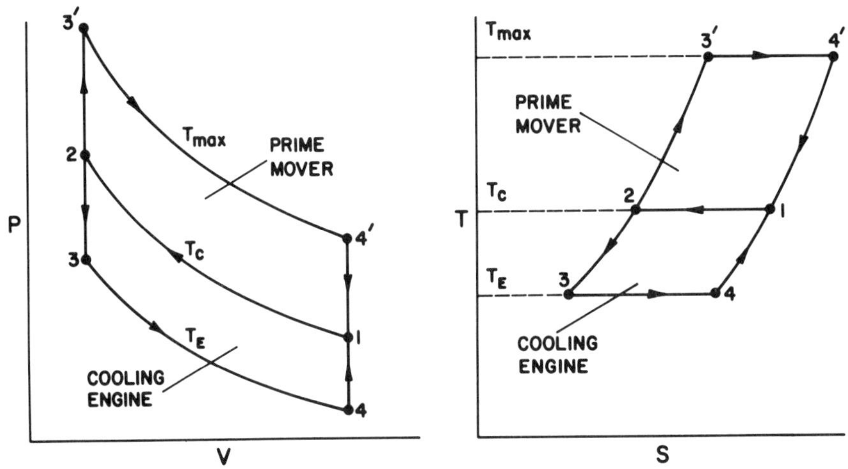

Fig. 4.4 Stirling cycle as a cooling engine and as a prime mover. In both applications the compression process occurs at ambient temperature T_C. The expansion process occurs at a low temperature T_E in the cooling engine and at a high temperature T_{max} in the prime mover. In both cases heat is absorbed during expansion and rejected during compression.

When the Stirling engine is used as a power system, heat at high temperature T_{max} is supplied to the cycle. Part of the heat is converted to work, and part is rejected as heat at ambient temperature T_C. The machinery is precisely the same as a refrigerator. Comparison of the cycles for operation as a power system and refrigerating machine is given in Fig. 4.4. In operating as a power system the pressure and temperature are greater in expansion than in compression. In operating as a refrigerator the reverse is true: the pressure and temperature are both less in expansion than in compression. A Stirling–Stirling, sometimes called a *duplex Stirling* system, is one where a Stirling power system generates sufficient work to drive a Stirling refrigerator. Very compact arrangements can be devised with the same working fluid at a common mean pressure and, quite often, a common piston.

As a heat pump the Stirling cycle operates exactly as the refrigerator with the temperature of the expansion space T_E less than the compression space T_C. The difference between operation as a heat pump and refrigerator is that in the heat pump both T_E and T_C are increased. Whereas in both the power system and refrigerator application, T_C is customarily the ambient atmospheric temperature at which cooling water is available, for a heat pump, T_C is the superambient temperature at

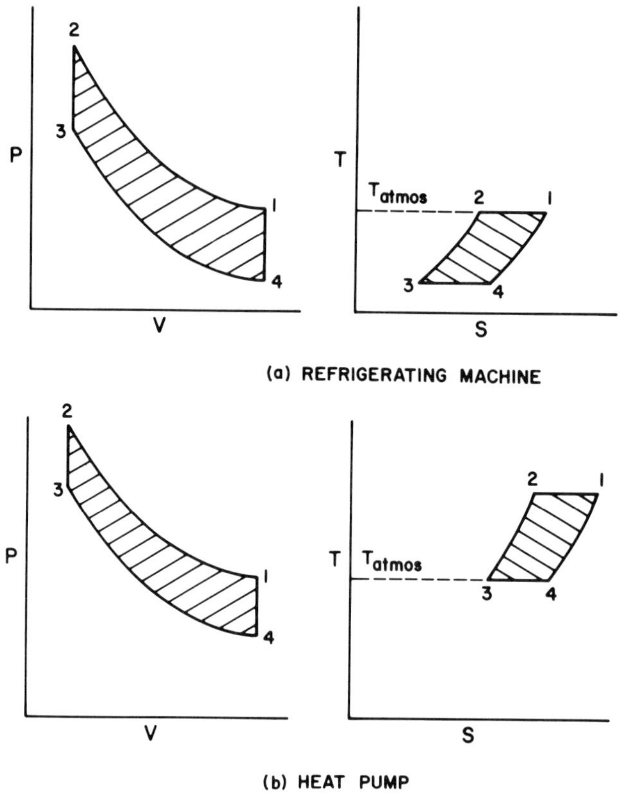

Fig. 4.5 Stirling cycle as a heat pump and as a cooling engine. When operating s a cooling engine or as a heat pump the expansion process occurs at a temperature less than that of the compression process and an input of work to operate the cycle is required. The cooling engine absorbs heat during expansion at temperatures less than the ambient and rejects heat at near-ambient temperatures. The heat pump absorbs heat during expansion at ambient temperature and rejects heat at temperatures greater than ambient.

which heat is rejected from the system. The heat rejected is the useful output for heating a concert hall or office building. Heat is supplied to the cycle (at T_E) from atmospheric air, river water, or ground source at approximately ambient atmospheric temperature and an input of work is necessary to drive the system.

A comparison of the Stirling cycle as a heat pump and a refrigerator is drawn in Fig. 4.5. In both cases, work from an external source is required to drive the cycle and is equivalent to area (1–2–3–4). In the case of the heat pump the useful product is the heat rejected at temperature T_C and the performance of a heat pump is therefore assessed as

$$(COP)_{HP} = \frac{\text{heat rejected}}{\text{work done}} = \frac{T_C}{T_C - T_E} \tag{4.6}$$

which is the inverse of the thermal efficiency.

The coefficient of the performance of a refrigerator

$$(COP)_{ref} = \frac{T_E}{T_C - T_E} \tag{4.7}$$

is not the inverse of thermal efficiency.

4.7.3 A versatile demonstration engine

In the late 1950s when the Philips Stirling engine air liquefiers were introduced it was common for them to demonstrate the versatility of the Stirling engine for operation as a power system, cooling engine, or a heat pump. A simplified diagram of the engine used for the demonstration is shown in Fig. 4.6. The engine was of a type known as the

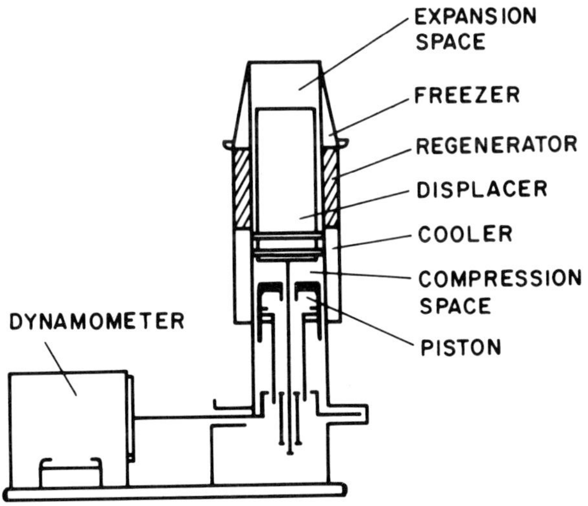

Fig. 4.6 Diagram of the versatile prime mover/refrigerator/heat pump Stirling engine demonstrated by Philips.

piston-displacer machine rather than the opposed piston version used above to describe the cycle.

In this machine the space above the displacer, below the cylinder head, is the expansion space. Between the underside of the displacer and the top of the piston is the compression space. The two spaces are in communication through three heat exchangers, a water-cooled cooler, a regenerator, and a finned freezer. The machine was mounted on a bed-plate and coupled to an electric dynamometer capable of operating as a motor or as a generator in either direction of rotation. Helium or hydrogen gas was used as the working fluid. A thermocouple was attached to the cylinder head and a large temperature scale was projected on to the wall of the lecture room with a moving light spot to indicate the cylinder head temperature.

The engine would be started initially as a refrigerator with the dynamometer working as a motor driving the machine in the forward direction. The cylinder head around the expansion space would cool rapidly and, on the temperature scale, one could observe the decrease in temperature of the cylinder head. At about 80 K air condensed on the cylinder head trickled into a collector at the base of the head. After a short time power to the motor would be switched off and, in less than a revolution, the machine would stop and self-start in the reverse direction.

It was now operating as a power system generating power. With the reversal of rotation the expansion and compression spaces switched locations so the expansion space was now between the displacer and piston. Heat was drawn from the cooling water to drive the engine and rejected from the cylinder head to the pool of liquid air serving as the engine coolant. Of course the reservoir of liquid air was rapidly dissipated and soon the temperature of the cylinder head started increasing. The engine continued to run but with progressively less vigour as the temperature increased until finally it slowed to low speed.

At that point power would once more be supplied to the dynamometer causing it to run again as a motor but preserving the reverse direction of rotation. With this power input the temperature of the cylinder head continued to increase until eventually it attained and then exceeded the ambient temperature. The machine was now operating as a heat pump drawing heat from the cooling water and rejecting it at higher temperature to the surroundings of the cylinder head, now acting as the compression space. The temperature of the head would increase rapidly, whereupon the lighting in the room would be lowered and the cylinder head could be observed glowing cherry red. At that point, Jan Kohler, the Dutch father of the Stirling cryocooler and consummate showman, would break his discourse, reach into his pocket, withdraw a cigar and casually lean down to light it on the hot cylinder head. It was beautifully executed and done so matter of factly. Following this, power to the motor would be switched off and again, immediately the engine would stop and start running in the opposite direction.

With this second reversal the direction of rotation was now back in the forward direction as it had been initially. The expansion and compression spaces were restored to the configuration shown in Fig. 4.6. The engine now ran again as a power system, withdrawing heat from the hot cylinder head, rejecting it to the cooling water. After some time the cylinder head cooled and, eventually, there was insufficient temperature difference between the expansion and compression spaces and the engine stopped.

It was a remarkable demonstration, beyond the capability of any other machine. More recently the Western Research Center of the General Pneumatics Corporation of Scottsdale, Arizona, has also demonstrated a Stirling engine working in the different modes. This unit was developed as a spacecraft refrigerator for NASA using air as the working fluid. This feature was attractive to NASA because it used the spacecraft atmosphere and not the customary Freon. Use of the spacecraft atmosphere reduced the spacecraft fluid inventory and eliminated the potential hazard to astronauts posed by a leakage of the refrigerant. Given public concern about the attack on the Earth's ozone layer by Freon in the atmosphere the machine has obvious potential for Freon-free refrigeration. The machine also operates effectively as a power system when heated by gas combustion or electric heat and so would be an effective teaching tool for schools and university departments.

The Philips machine operated with compressed hydrogen and was never put into general production because of concerns about its safe operation in unskilled hands. The General Pneumatics machine uses air at low pressure and so is safe to use. Under these conditions it cannot achieve liquid air temperatures but nevertheless cools sufficiently quickly to establish a significant water vapour frosting on the cylinder head that is sufficient to demonstrate the Stirling engine capability to function as a refrigerator.

The versatility for the same unit to operate as a prime mover, refrigerator, and heat pump has practical implication for the future. A duplex Stirling unit, otherwise known as a Stirling–Stirling unit, wherein a Stirling refrigerator is driven by a Stirling prime mover, has already been demonstrated by William Beale at Sunpower Inc. The use of a common working fluid has outstanding attractions to overcome seal problems. The combination may be a solar, isotope, or combustion heated cryocooler. More likely, in the near future, it will be developed as a gas-fired air conditioning refrigerator.

Another possibility in development at the General Pneumatics Corporation and the University of Calgary is a natural gas-fired Stirling–Stirling natural gas liquefier. One application for this unit is to liquefy the pipeline gas delivered to homes, farms, and fleet gas depots for use as an automotive or farm vehicle fuel. Another application, in a much larger size, is to liquefy well head gas in the thousands of 'capped' gas wells located in many parts of the world remote from a pipeline with not enough gas to justify the link. Following expansion from the well head pressure the raw gas would be cooled in a switching recuperative heat exchanger. The propane, butane, and other organic 'heavy ends' in the gas would be precipitated on the walls of the heat exchanger. After a suitable interval the flows would be switched. The non-condensables leaving the liquefier, namely nitrogen, would now pass through the contaminated channels where the precipitated heavy ends would re-evaporate and pass to the combustion chamber. In this way much of the heat necessary to drive the liquefier would be derived from the 'waste' product of the well head gas.

The Stirling–Stirling or duplex Stirling refrigerator should not be confused with the Vuilleumier cycle machines which have only displacers, both hot and cold. This is a much simpler arrangement that is, at once, less effective and more costly to build and operate.

Another concept to utilize the versatility of the Stirling engine is to use it as a power system, extracting heat from a thermal battery (lithium fluoride) as an automotive propulsion engine. For downhill operation, or when braking, the engine could be converted to function as a heat pump feeding thermal energy into the thermal battery. In practice it is not necessary to physically reverse the engine. The same effect can be gained in some arrangements simply by switching fluid connections. The combined regenerative braking/propulsion concept is particularly attractive in mining applications where trucks are used on loaded downhaul operation and for underground mines and also for railroad locomotives. Both the mining truck and railroad locomotive units could operate just as effectively with coal combustion as with the diesel oil used for the present internal combustion diesel units. The same concept will likely become attractive for road vehicles as the fossil fuel supplies dwindle and the world moves into the 'electric economy'. The Stirling engine with thermal storage will likely become the preferred automotive propulsion unit. On arrival at work or at home commuters will plug in their car for thermal recharge provided by electric heat or natural gas combustion. This commuter behaviour prototype already exists in Canada where in winter it is customary when parking to 'plug in' an engine block heater to keep the engine warm enough to start easily.

A. J. Oxley of the UK Rutherford Laboratory has explored the use of liquid nitrogen as the coolant for Stirling engines with ambient temperature 'heating'. Engines could be energized from an ambient temperature source to operate as a prime mover for underground mining applications with no emission of noxious fumes. Other possibilities for large-scale power generation exist using the temperature difference between sea water and the ship loads of liquid natural gas now coming into the United States, Japan, and Europe on a regular scheduled basis. Currently the gas is vaporized using sea water 'heating' for insertion into gas transmission distribution pipelines but could equally well be utilized to generate power and so recover some of the millions of horsepower hours consumed in liquefaction of a shipload of liquid natural gas.

4.7.4 Theoretical analysis of Stirling cryocoolers

The classical theory for Stirling engines, the analysis for isothermal compression and expansion developed by Gustav Schmidt (1861), has been restated in many recent books on Stirling machines, such as Walker (1983).

More advanced analysis of Stirling machines usually involves the use of computers. Walker *et al.* (1993) has briefly reviewed the status of computer simulation of Stirling cryocoolers and power systems including the commercial availability of CRYOWEISS, a Stirling cryocooler design aid and MARWEISS, a Stirling power system design aid both developed at the University of Calgary, for general use with microprocessors using MS-DOS 2.0 or later.

4.7.5 Commercial applications of Stirling cryocoolers

Miniature Stirling cryocoolers are used in large numbers (thousands a month) to cool the infrared detectors of night vision equipment and missile guidance systems. This application is principally military but there is increasing use for a variety of civil, scientific, and medical infrared imaging applications. Figure 4.7 shows one form of Stirling engine used for this purpose, an integral two-piston arrangement. Figure 4.8 shows another type of Stirling cryocooler in general use for infrared applications. This is a split-Stirling in which the cold expansion section is physically separated from the larger compression section which includes the drive motor. The two elements are coupled by a single fine-bore transfer tube.

The split-Stirling cryocooler is particularly useful for gimballed sights. In this application a low mass of the infrared detector apparatus is desired to facilitate rapid, smooth movements of the sight with low acceleration forces.

(a)

FLIR MODULAR COOLER

(b)

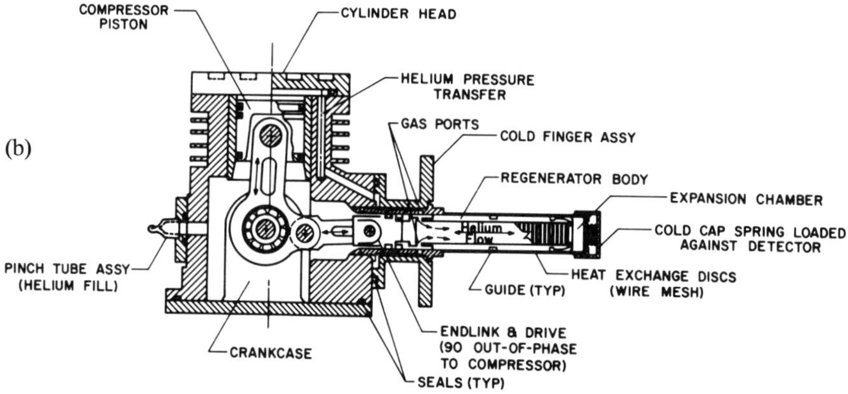

Fig. 4.7 Small integral-Stirling cryocooler for forward-looking infrared application.

The technology for miniature Stirling cryocoolers has recently been reviewed extensively by Walker (1989) including a brief description of the new generation of microminiature Stirling cryocoolers with power inputs of only 3 or 4 watts developed by the Israeli company Ricor for use in battery powered, hand-held, infrared imaging equipment (represented in the US by Cincinnati Electronics).

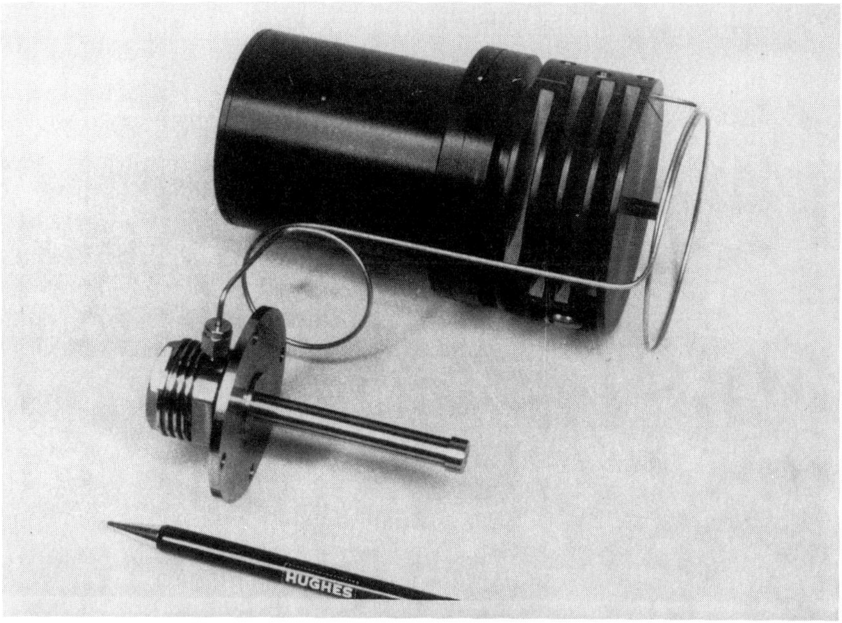

Fig. 4.8 Split-Stirling cryocooler for forward-looking infrared application.

Another excellent review of the technology of small Stirling engines was given by Berry (1993) with important details of current developments in linear-coolers with advanced demand control features.

Larger versions of Stirling cryocoolers are also available. Figure 4.9 shows a Stirling air liquefier manufactured by the Cryogenics Division of the Philips Electrical Corporation, Eindhoven, Holland†. This produces 5 or 6 litres of liquid air per hour. It is available in the single-cylinder version shown and in a larger four-cylinder version with correspondingly greater capacity. A range of auxiliary equipment is available including an air separation column to separate the liquid air into liquid nitrogen (safer to use than liquid air) and gaseous oxygen.

A version of the air liquefier but with two stages of expansion is also available. This is capable of achieving cryogenic refrigeration at two levels, 80 K and a lower temperature with the minimum ranging down to 20 K.

A much larger Stirling cryocooler, capable of 20 kW of cooling at 110 K was developed by Philips and is shown in Fig. 4.10. The large

*Now the General Stirling Co. Ltd.

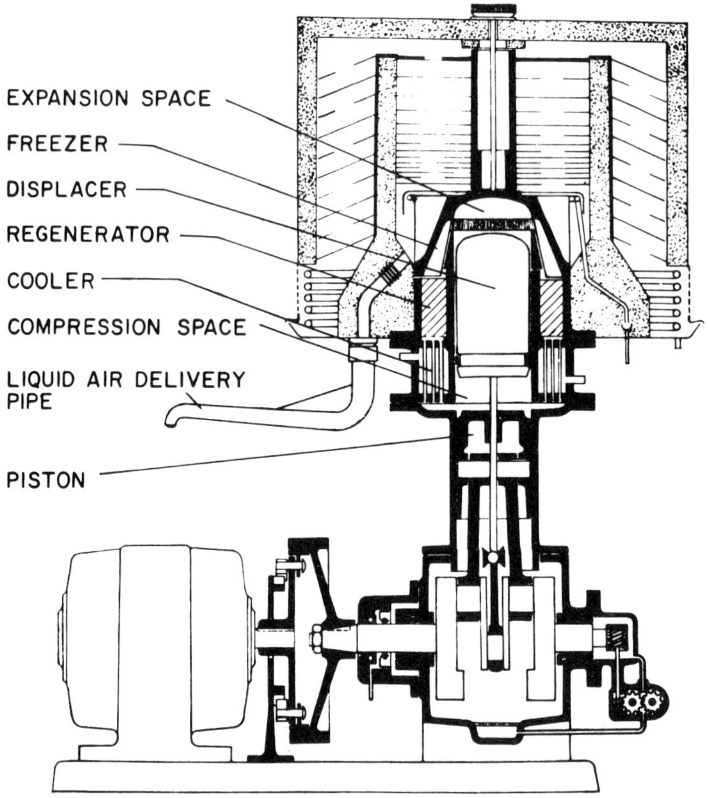

Fig. 4.9 Philips Stirling air liquefier.

unit was intended to reliquefy the vapour boil-off from liquid natural gas (LNG) storage tanks both stationary and on board LNG-carrying ships. It functioned very satisfactorily but the capital cost was too high ($100,000 in the early 1970s) for the unit to go into production. Further details of these larger units including the original references were given by Walker (1983).

Several Stirling cryocoolers acting as nitrogen vapour liquefiers are being developed at the present time (1993). One intended application was to use these liquefiers to recondense the nitrogen vapour generated by supercomputers operating with their VLSI processing boards submerged in liquid nitrogen. Typically the cooling load for a single computer board was quoted as 700 watts at 80 K approximately but computers with multiple units were being manufactured. The largest known to the

Fig. 4.10 Large Philips Stirling cryocooler.

authors called for the use of 34 boards and involved a cooling load of $34 \times 700 = 24$ kW at 80 K, already in excess of the capacity of the large Stirling cryocooler shown in Fig. 4.10. It was thought likely that for these higher loads a Claude cycle cryocooler with high-speed turbo-expanders would be favoured over the larger, slow-running Stirling cryocoolers. Unfortunately, the development of supercomputers operating at cryogenic temperatures now appears to be 'on hold' and it is not clear to the lay observer when and if such development will be resumed. At the other end of the scale efforts to develop even smaller Stirling cryocoolers for cryogenic sensors and cold electronics are continuing. There are some who predict the era of cold electronics systems will result in changes as profound as the invention of the transistor. However there is general agreement that the development of miniature and microminiature low-cost, disposable, safe, reliable cryocoolers that anyone can use simply by operating a switch is a necessary precursor to the widespread use of cold electronic systems.

Semiconductor electronics systems work progressively faster with reduced 'noise' as the operating temperature is reduced. There is little to be gained, however, by operation at temperatures much below 80 K.

Superconducting electronics require cooling to cryogenic temperatures, to some level below the maximum superconducting transition temperature of the material. The recent discoveries that certain copper oxides and other materials become superconducting at temperatures (100 K and more) well above the transition temperature of well-established superconducting materials (for example niobium at 20 K) offers exciting possibilities for future superconducting electronics systems at 80 K or thereabouts.

This congruence of interest of semiconductor and superconducting electronics at 80 K, a cryogenic temperature level relatively easy to obtain with a single stage of expansion, is precisely the region where the greatest experience has been gained in the development of miniature Stirling coolers for infrared night vision and missile guidance systems. Such units presently cost more than their weight in gold and have limited life but significant developments are in progress. Walker and Bingham (1991) recently discussed the possibilities for applying silicon technology used for semiconductor VLSI systems to microminiature Stirling cryocoolers.

One form of Stirling cryocooler currently attracting much attention is the so-called 'Oxford derivative' machine. This is a split-Stirling unit with a linear electric motor drive to the compressor and a linear electric generator on the displacer unit. A substantial electronics package is associated with the cooler to control and regulate the operation of the compressor and displacer unit as well as maintain the proper phase relationship. The reciprocating members in both the compressor and displacer unit are mounted with flexure diaphragms or springs so there are no rubbing seals. This is held to endow the machine with the potential for long life and so is currently much favoured for aerospace application where 5–10 year operating lifetimes are sought. The original unit of this type was developed at Oxford University (Davey 1990) and was further developed in England by British Aerospace, Lucas Aerospace, and the Science Research Council, Rutherford Laboratories. Its potential for US aerospace applications was perceived by NASA Goddard personnel who have enthusiastically encouraged further development. Development programs of 'Oxford type' coolers currently include a joint British Aerospace/Ball effort, a joint Lucas/Lockheed effort and a separate program by Hughes Aircraft. Recent progress on these programs was reviewed by Sparr *et al.* (1993).

Careful evaluation tests of long-life cryocoolers are being carried out at the Jet Propulsion Laboratory and the US Air Force Space Defence Initiative Organisation 'Brilliant Eyes' program (Ross 1993) including evaluation and life tests of Oxford type cryocoolers. Details of the JPL Test Facility and characterization program were given by Ross and Johnson (1993), Johnson *et al.* (1993), and by Glaser *et al.* (1993). A

review of the life-testing program for Oxford type cryocoolers intended for use with European Space Agency spacecraft was given recently by Jewell *et al.* (1993).

4.8 Gifford–McMahon cryocoolers

4.8.1 Cycle of operation

The elements of a Gifford–McMahon cryocooler are shown schematically in Fig. 4.11. They consist first of a cylinder, four or five diameters long, closed at both ends and containing a displacer made of material having a low thermal conductivity and about three-quarters of the cylinder in length. The displacer is connected to a light drive mechanism so that it can be moved up and down in the cylinder. As the displacer reciprocates in the cylinder the two volumes, one above and one below the displacer, vary cyclically from zero to maximum but the total volume remains constant.

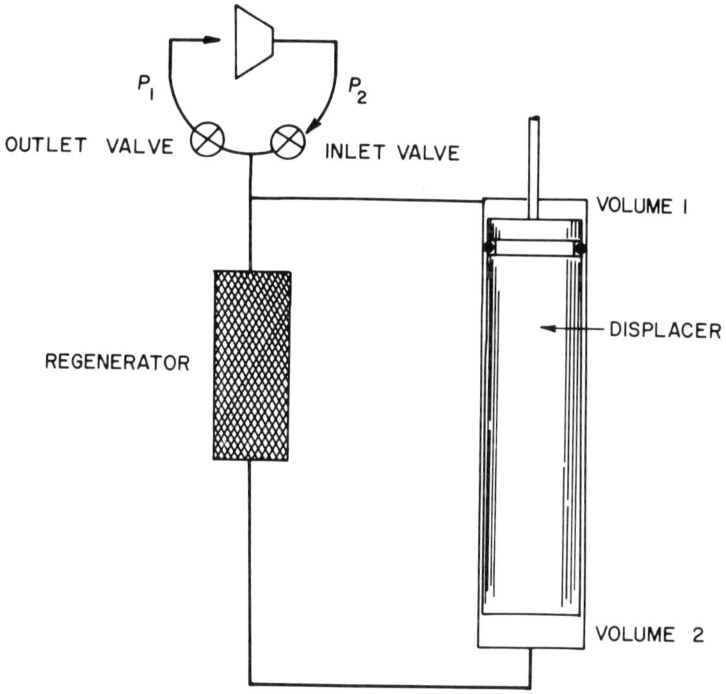

Fig. 4.11 Elements of a Gifford–McMahon cryocooler.

The two spaces are connected through a thermal regenerator and to a gas supply. The gas supply system consists of inlet and outlet valves, a gas compressor, and both high- and low-pressure reservoirs. The valves are coupled to the drive mechanism so their operation is synchronized with the position of the displacer. A heat exchanger and a gas cleaner, or filter (neither shown in Fig. 4.11), are included downstream of the gas compressor to clean and cool the gas to ambient temperature after compression.

It is possible that all these elements could be combined into a single integrated unit. In practice, however, Gifford–McMahon cryocoolers are invariably split systems with the cold expansion part isolated as required from the hot compressive part but coupled through the high-pressure inlet and low-pressure ducts.

The pressure above and below the displacer is the same except for small pressure drops across the regenerator when gas is flowing through it. Therefore, virtually no work is required to move the displacer in the cylinder apart from minor mechanical friction losses due to the rubbing seal at the upper end of the displacer. As the displacer moves no work is done on the gas and the gas does no work on the displacer. Gas flow to and from the displacer cylinder is controlled by operation of the inlet or outlet valves.

The displacer is a loose fit in the cylinder except at the top where it is equipped with a dynamic (sliding) seal to prevent leakage past the displacer from one space to the other. The pressure is virtually the same in both spaces so the seal problem is not severe and, furthermore, some leakage can be tolerated between the two spaces.

The regenerator consists of finely divided metallic material which cools the gas when passing downwards to the cold space and heats the gas when passing upwards from the cold space.

The engine operates as follows:

1. *Pressure build-up*. With the displacer at the bottom of the cylinder and exhaust valve closed the inlet valve is opened, increasing the pressure level in the system. Fluid enters through the inlet valve to fill the regenerator and the space above the displacer, volume 1.

2. *Intake stroke*. With the inlet valve open the displacer is moved from the bottom of the cylinder to the top. This displaces high-pressure fluid from the space above the displacer, volume 1, through the regenerator to the space below the displacer, volume 2. In passing through the regenerator the gas cools causing the pressure to decrease and further gas to enter the system to maintain the maximum cycle pressure.

3. *Pressure release and expansion*. With the displacer at the top of the cylinder the inlet valve closes and exhaust opens. Fluid escapes and the pressure falls. The drop in pressure corresponds to an expansion of the gas causing its temperature to fall. The decrease in temperature of gas in the bottom cylinder space, volume 2, causes heat to be abstracted from the surrounds of the cylinder, a process which is the useful refrigeration process of the cycle.

4. *Exhaust stroke*. With the exhaust valve open, the displacer moves from the top to the bottom of the cylinder, displacing fluid from volume 22 below, to volume 1, above, the displacer. As it flows through the regenerator the fluid is heated by the matrix to near ambient temperatures. This process cools the matrix ready for the gas entering in the succeeding cycle.

During pressure build-up and expansion, the regenerator is most effective if gas can be made to flow through it at a constant rate. This will be achieved by progressive opening of the inlet and exhaust valves when the displacer is stationary at the bottom and at the top of the cylinder, respectively. The inlet valve opening should be extended over about 75 ° and the exhaust valve over about 110 ° of crank rotation.

Figure 4.12 shows the work diagrams for volumes 1 and 2 and Fig. 4.13 shows the temperature–time characteristic for gas in the engine. At the beginning of a cycle volume 1 is maximum and volume 2 is minimum. The pressure is low, represented on Fig. 4.12 and 4.13 by state (a).

The inlet valve opens and fluid enters volume 1 and the regenerator. The fluid contained therein is compressed and its temperature is consequently increased. It mixes with incoming fluid to a final temperature

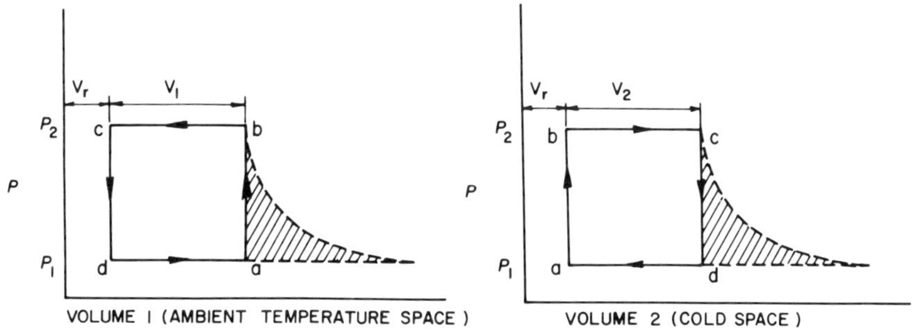

Fig. 4.12 Work diagrams for the Gifford–McMahon cryocooler.

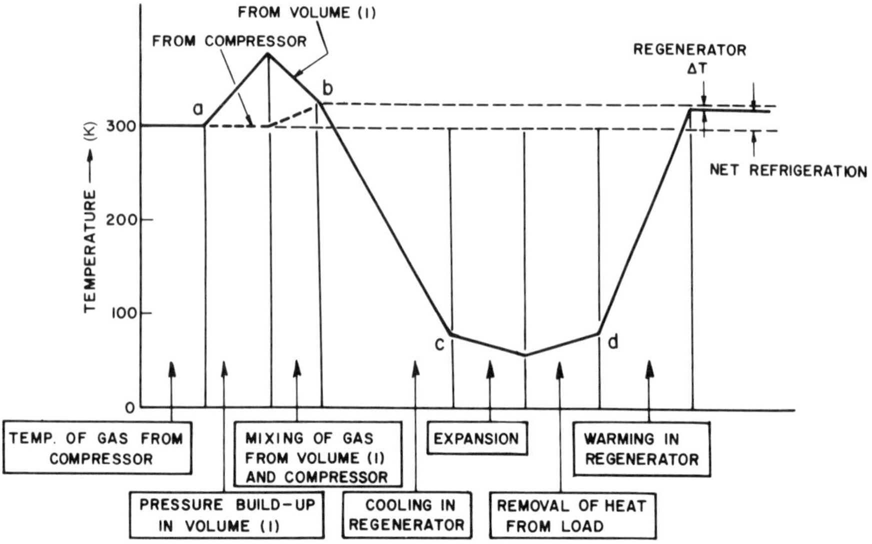

Fig. 4.13 Temperature–time history of gas in the Gifford–McMahon cryocooler.

of mixed fluid shown by state (b) on Fig. 4.13. Now the displacer moves from the bottom to the top of the cylinder and fluid is displaced from volume 1 to volume 2 through the regenerator. In passing through the regenerator the fluid cools to temperature (c) and more fluid enters through the inlet valve to maintain the pressure constant.

At (c) the inlet valve closes quickly and the exhaust valve opens slowly and progressively. The pressure falls as fluid leaves the engine through the exhaust valve. Fluid in volume 2 expands as the pressure decreases and the temperature of gas in volume 2 decreases. Heat transfer from the cylinder walls increases the gas temperature to (d) in Fig. 4.13.

Finally, the displacer moves from the top to the bottom of the cylinder causing fluid to return back through the regenerator to volume 1 or exhaust. In passing through the regenerator the temperature of the gas increases to a value slightly less than the mixture following admission of high-pressure gas.

The gas leaving the exhaust is, therefore, warmer than the intake feed gas. The heat gained in this way is, in fact, equal to the total refrigeration achieved by gas expanding in the cold lower region of the cylinder.

The net result of operating the system and investing a considerable input of work to drive the compressor is that heat is pumped from the low temperature at which it is abstracted from the surrounds of the displacer cylinder to above the ambient level at which it is rejected from the system. The useful product is the refrigeration effect generated during expansion.

The work done by the gas on the displacer is shown by the work diagram for volume 2. The work done by the displacer on the gas is shown by the work diagram for volume 1. The pressures are the same in both spaces so the areas of the work diagrams are identical (neglecting the displacer rod area). For an ideal gas and adiabatic conditions the area of the diagram for volume 2 is equal to the change in enthalpy of the gas in volume 2. It represents the net refrigeration available. Similarly, the area of the diagram for volume 1 represents the change in enthalpy of a perfect gas under adiabatic conditions.

4.8.2 *Fluid-driven displacer*

A refinement to the original Gifford–McMahon cryocooler occurred with the development of a technique known as 'cryogenic gas balancing'. The displacer mechanical drive was eliminated and a free displacer used instead. Movement of the displacer was caused by differential fluidic (gas pressure) forces acting on the displacer arising from the inlet and exhaust of gas through a simple, rotary cored, valve.

A schematic diagram of a single-stage Gifford–McMahon cryocooler with fluid-driven free displacer is shown in Fig. 4.14. It consists of two connected coaxial cylinders of different length and diameter containing the free displacer. The displacer is about three-quarters of the length of the larger cylinder and has a coaxial extension to fit the smaller cylinder. This creates three chambers of variable volume depending on the position of the displacer.

Chamber 1 is the volume between the top of the displacer extension and the small cylinder head. Chamber 2 is the volume between the top of the displacer and the large cylinder. Chamber 3 is the volume between the bottom of the displacer and displacer extension to prevent the leakage of gas past the displacer from one chamber to another.

Chambers 2 and 3 are interconnected by a thermal regenerator of finely divided metallic material. The pressure in both is, therefore, the same, apart from small differences due to the pressure drop in the regenerator matrix when gas is flowing through. Gas is admitted and exhausted from chamber 1 and the regenerator conduit through a rotary cored valve. The same valve regulates both the inlet of high-pressure gas from the compressor and the exhaust of low-pressure gas.

The sequence of operations illustrated in Fig. 4.15 is as follows:

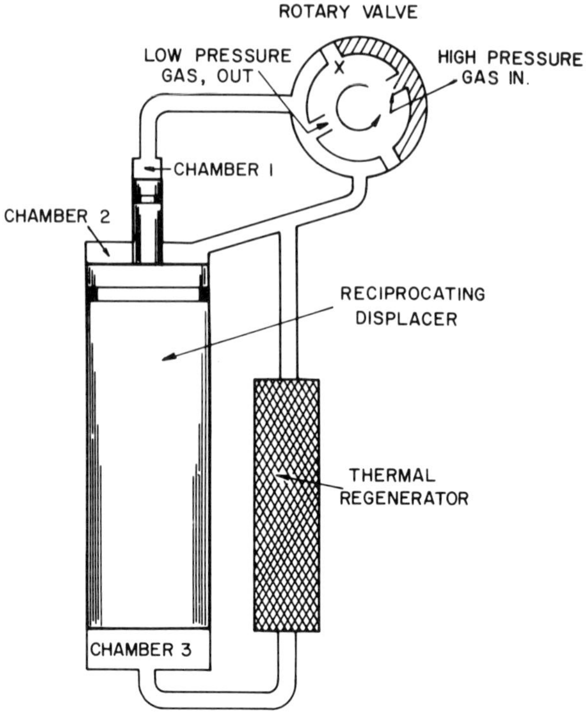

Fig. 4.14 Schematic diagram of Gifford–McMahon cryocooler with a free fluid-driven cryocooler.

1. *Pressure build-up*. At the beginning of the cycle the displacer is at the bottom of the cylinder so the volume of chamber 3 is zero and the volumes of chambers 1 and 2 are at their maximum value. The rotary valve is in such a position that chamber 1 is full of gas at a high pressure and chambers 2 and 3 are connected to exhaust and at low pressure. This is represented in Fig. 4.15(a). When the valve has turned to connect chambers 2 and 3 to the inlet port the pressure increases as represented in diagram (b). Fluid enters through the inlet port to fill chamber 2, the regenerator, and the clearance space in chamber 3.

2. *Intake stroke*. Further rotation of the valve opens chamber 1 to exhaust and the pressure decreases. The upward force on the displacer due to pressure in chamber 3 now exceeds the downward force due to pressure in chambers 1 and 2 by an amount equal to the pressure times the area of chamber 1. As a result, the displacer moves to

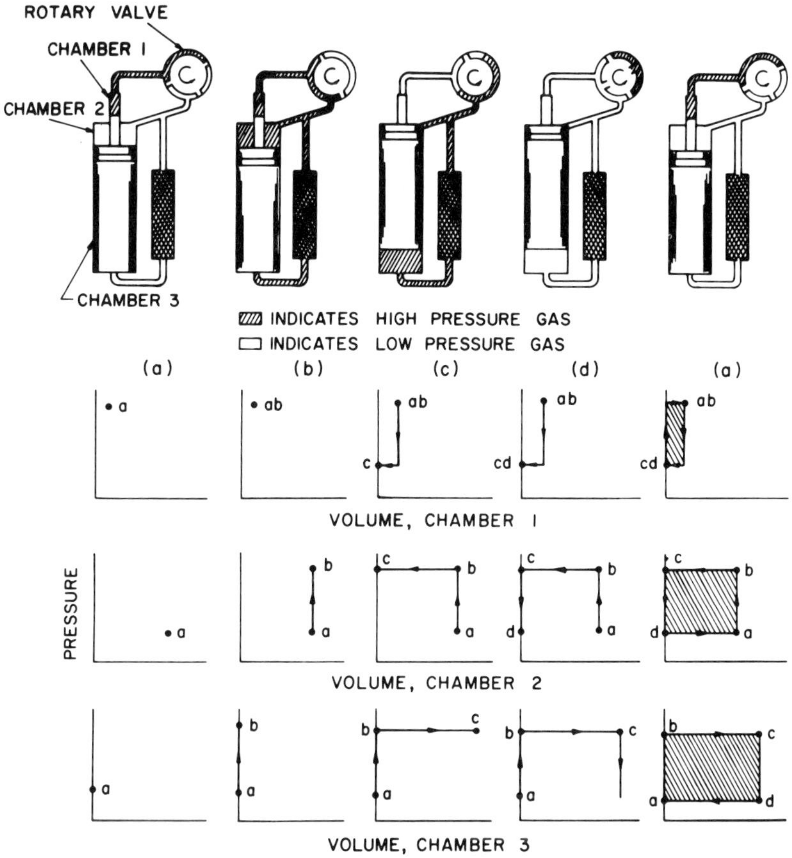

Fig. 4.15 Cycle of operation of the Gifford–McMahon cryocooler with free fluid-driven displacer.

the top of the cylinder (Fig. 4.15(c)) displacing fluid from chamber 2 through the regenerator to chamber 3. In passing through the regenerator the fluid cools and additional fluid enters to fill chamber 3 and the regenerator.

3. *Pressure release or expansion*. Further rotation of the valve opens chambers 2 and 3 to exhaust and the pressure decreases. Fluid flows out of chamber 3 through the regenerator during the expansion and the useful refrigerating effect is developed in chamber 3. The situation is represented in Fig. 4.15(d).

4. *Exhaust stroke*. The same low pressure now exists in all three chambers so the displacer will fall under gravity to the bottom of

the cylinder. Should seal friction prevent this the valve soon opens chamber 1 to the high-pressure gas supply. The pressure force acting downwards then moves the displacer to the bottom of the cylinder thereby displacing fluid from chamber 3 to chamber 2. In passing through the regenerator the fluid is heated by the matrix to near ambient temperature. This process removes heat from the matrix, cooling it in preparation for the gas entering in the next cycle. The increase in temperature of the gas leaving decreases the density of the gas so that much of the fluid leaving chamber 3 passes to exhaust rather than remaining in chamber 2.

The work diagrams for the unit are similar to those discussed earlier for the engine with a mechanically driven displacer. The small diagram for chamber 1 represents the pneumatic work done in driving the displacer up and down the cylinder. The diagram for chamber 3 represents the change in enthalpy and, for adiabatic conditions with an ideal gas, the refrigeration effect of the process per cycle. The diagram for chamber 2 is virtually identical but reversed in direction to that for chamber 3. A temperature–time history of gas in the engine would be the same as that presented earlier.

4.8.3 System optimization

Some of the interesting heat transfer fluid friction and mechanical design problems of Gifford–McMahon refrigerating engines were discussed by Gifford and Hoffman (1961). An important factor is the valve action and motion of the displacer. For best performance of the regenerator the gas flow should be constant one way for half the cycle and constant the other way for the remainder of the cycle. Slight variations are permissible but very large surges of fluid through the matrix seriously affect regenerator effectiveness.

These requirements are not compatible with the flow patterns necessary to maximize the refrigeration effect in the expansion space for a displacer moving with continuous harmonic motion. To accomplish that sudden changes in pressure from one extreme to the other are required at a time when the displacer is stationary at the ends of its stroke. In order to maximize the area of the work diagram for the expansion space a rapid *increase* in pressure is required when the displacer is at the *bottom* of the cylinder and correspondingly a sudden *decrease* when the displacer is at the *top* of the cylinder. Any arrangements made to achieve this would result in periodic rapid surges of gas through the system and lead to a serious decline in regenerator performance.

An effective solution to the problem was achieved by:

(1) causing the displacer to move in a discontinuous manner so that there are lengthy pauses at each end of the stroke, and

(2) increasing and decreasing the pressure through valves which open slowly to bleed gas in and out at an approximately constant rate.

4.8.4 Regenerative displacer

The regenerator of the system in Fig. 4.11 is shown to be separate from the displacer. In practical Gifford–McMahon cryocoolers the regenerator is nearly always contained within the displacer to effect a simplification in the design and mechanical arrangement of the machine.

4.8.5 Theoretical analysis

An elementary analysis of Gifford–McMahon refrigerators was included by Gifford (1966) in his disclosure of the new cycle. Later Ackerman and Gifford (1971) gave a heat balance analysis of an experimental Gifford–McMahon refrigerator. Thirumaleshwar and Subramanyam (1986) presented a more sophisticated theoretical model and followed this with several further papers about different aspects of their work. More recently Minas and Hualde (1992) presented the results of their dynamic modelling of a Gifford–McMahon cryorefrigerator. This paper is of interest in providing a good summary of the work, with principal references, in this field.

4.8.6 Compressors in Gifford–McMahon cryocoolers

The compressors used for many of the early Gifford–McMahon cryocoolers were frequently the same compressors used for refrigeration and air conditioning systems but adapted for helium service. These compressors were usually oil lubricated with no hermetic seal between the crankcase and the working space. Consequently substantial oil clean-up apparatus and equipment downstream of the compressor was necessary. The alternative approach of operating a non-oil-lubricated compressor was investigated but the dry rubbing seal and bearing materials available were not adequate for the long trouble free service desired.

Eventually the commercial market for Gifford–McMahon cryocoolers developed to the extent that compressor manufacturers became sufficiently interested to develop special compact compressors for the service.

4.8.7 Commercial applications of Gifford–McMahon cryocoolers

Gifford–McMahon cryocoolers are made by a dozen or more manufacturers worldwide in a variety of sizes, refrigeration capacities, and temperature levels, one, two; and three stages of expansion.

Table 4.1 Technical specifications for several Gifford–McMahon cryocoolers (after Cryomech Inc., Syracuse, NY).

Type of refrigerator	AL01	AL03	AL05
OUTPUT Cooling capacity (Watts)	1W@25K 3W@27K 20W@80K	 2W@30K 8W@80K	 50W@40K 120W@80K
Lowest temperature:	23K	19K	23K
INPUT			
1) VAC-Phase-Hz	115–1ϕ-60 240/220–1ϕ-50	115–1ϕ-60 240/220–1ϕ-50	230/208–3ϕ-60 415/380–3ϕ-50
2) Power required (kW)	1.2	0.7	4.5
3) Cooling water	2 litres/min.	(1 litres/min)	4 litres/min.
Refrigeration cycle:	Gifford–McMahon	Gifford–McMahon	Gifford–McMahon
Standard cooling (optional):	Water (Air)	Air (Water)	Water (Air)
Expander:	Single stage	Single stage	Single stage
Cool-down time:	15 min. to 80K	30 min. to 80K	15 min. to 80K
Revolutions/min:	144	144	144
Mean time between maintenance (hours): Warranty on all parts 3 yrs or 6,000 hrs	10,000	10,000	10,000
Weight:			
Cold head	5 lb (2.3 kg)	4 lb (1.8 kg)	25 lb (11.3 kg)
Compressor	125 lb (56.7 kg)	125 lb (56.7 kg)	150 lb (68.0 kg)
Applications:	Solid state detector cooling. Air liquefaction. Thermal shielding. UHV	Solid state detector cooling. Thermal shielding. UHV	Air liquefaction. Thermal shielding. UHV

GB220	GB04	GB07	GB37
1W@10K 4W@20K 4W@20K with 15W@80K	2W@10K 9W@20K 7W@20K with 60W@80K	10W@12K 15W@16K 20W@20K with 60W@80K	10W@11K 30W@20K 30W@20K with 50W@80K
7.5K	7.5K	8K	8K
230/208–1ϕ-60 240/220–1ϕ-50 2.0 2 litres/min.	230/208–3ϕ-60 415/380–3ϕ-50 4.5 4 litres/min.	230/208–3ϕ-60 200–3ϕ-50 10.5 8 litres/min.	230/208–3ϕ-60 200–3ϕ-50 10.5 8 litres/min.
Gifford– McMahon	Gifford– McMahon	Gifford– McMahon	Gifford– McMahon
Water (Air)	Water (Air)	Water	Water
Two stage	Two stage	Two stage	Two stage
40 min. to 10K	40 min. to 10K	40 min. to 10K	40 min. to 10K
144	144	144	144
10,000	10,000	10,000	10,000
8.5 lb (3.8 kg) 125 lb (56.7 kg)	25 lb (11.3 kg) 150 lb (68.0 kg)	35 lb (15.9 kg) 550 lb (249.5 kg)	50 lb (22.7 kg) 550 lb (249.5 kg)
Superconducting magnet cooling. Magnetometers. Cryopumping. Sample cooling. UHV	Superconducting magnet cooling. Magnetometers. Cryopumping. Sample cooling. UHV	Superconducting magnet cooling. Cryopumping. Sample cooling.	Superconducting magnet cooling. Cryopumping. Sample cooling.

Fig. 4.16 Gifford–McMahon cryocooler. CTI Model No. 0120 (courtesy Cryogenic Technology Inc.).

Table 4.1 summarizes the technical specifications for several Gifford–McMahon cryocoolers currently available from Cryomech Inc., Syracuse, New York, a company founded by William Gifford, one of the system inventors, when a professor at Syracuse University, New York.

Gifford–McMahon cryocoolers tend to be relatively expensive, relatively large, heavy, slow-running units of great reliability having maintenance intervals of 10,000 hours (once a year) or more. They have come to be used in applications where the ability to function for long periods without maintenance is prized above considerations of cost, size, and weight, typically commercial and industrial applications where low-capacity cryogenic refrigeration is routinely required in continuous or semi-continuous operation. Figure 4.16 shows a typical Gifford–McMahon cryocooler in widespread commercial application. They are widely used with two stages of expansion to achieve temperatures of 20 K for cryopumping in the semiconductor processing industry where high levels of vacuum are routinely and reliably required.

More recent developments include a two-stage Gifford–McMahon cryocooler capable of generating 0.5 watts of refrigeration at 4.2 K.

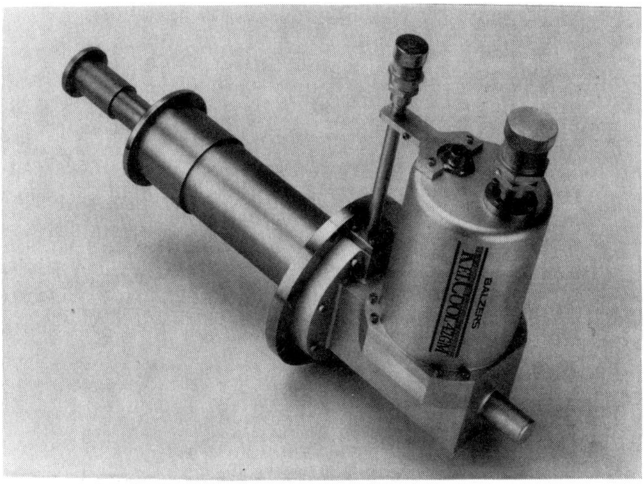

Fig. 4.17 Cold head of a two-stage Gifford–McMahon cryocooler capable of 0.5 W refrigeration at 4.2 K (courtesy of Balzers High Vacuum Products).

Figure 4.17 shows the cold head of this machine introduced in 1992 by Balzers High Vacuum Products. This machine uses unconventional 'rare earth' material for the second stage matrix. Recent cryocooler conferences (fifth, sixth, and seventh International Cryocooler Conferences) have included papers, principally from Japanese sources, about temperatures of 3.5 K and up achieved with two and three stage Gifford–McMahon refrigerators using rare earth matrices for the low temperature regenerators.

4.9 Solvay cryocoolers

A Solvay cryocooler is very similar to a Gifford–McMahon cryocooler with the important difference that a Solvay machine uses a piston instead of a displacer. The use of a piston instead of a displacer results in the production of net work output by the piston from the expansion of the gas in the cold expansion space. This work output increases the effectiveness of the expansion thereby increasing the refrigerating capacity of a given size of unit. Moreover, the input of work per unit of refrigeration (watts input power per watt refrigeration) is decreased. Detailed consideration of the operating cycle was given by Walker (1983).

The disadvantage of the Solvay machine is that sealing of the piston is more complicated and difficult to achieve than a displacer.

According to Collins and Cannaday (1958) the inventor, E. Solvay (1887), was never successful in achieving cryogenic temperatures with his prototype unit. The concept was moribund until its rediscovery by William Gifford and Howard McMahon in the late 1950s more or less simultaneously with the no-expansion work Gifford–McMahon machine described above. Small experimental prototype Solvay cryocoolers were made but the principal development effort was devoted to the displacer type machine that now carries their name. Parenthetically it is noteworthy that the displacer machine was also a reinvention of a refrigerator originally due to Davy Postle of Stanthorpe, Queensland (Thomas 1873).

4.9.1 Commercial applications of Solvay cryocoolers

Cryocoolers described as operating on the modified Solvay cycle have been in production for many years at the AP Division of Intermagnetics Inc. of Allentown, Pennsylvania (formerly the Air Products and Chemicals Cryogenic Division). These units, called Displex refrigerators, and described by Longsworth (1971) utilize a displacer (not a piston) in the expansion unit. However, the displacer is coupled to a fluid damping dashpot with high frictional dissipation of energy. The dashpot restrains the motion of the displacer creating the same effect of gaseous expansion that would be achieved by a piston producing output work by expansion of the gas. These units have been successfully utilized in the same type of application as the Gifford–McMahon machines described above.

4.10 Vuilleumier cryocoolers

A Vuilleumier, or VM, cryocooler shown in Fig. 4.18 uses an input of thermal energy at high temperatures to generate cyclic pressure fluctuations of the working fluid contained in the closed volume of the unit.

The pressure variations are produced by the action of a reciprocating displacer shuttling the working fluid gas periodically from an ambient temperature space to a high-temperature space through a regenerator. This is sometimes described as a thermal compressor and compares with a Stirling cryocooler having a mechanical compressor.

In the Vuilleumier machine there is a second smaller reciprocating displacer in the cold cylinder. This corresponds exactly with the displacer found in Stirling cryocoolers. By the action of this reciprocating displacer working fluid is shuttled back and forth through a regenerator between a space at the ambient temperature and a cold region.

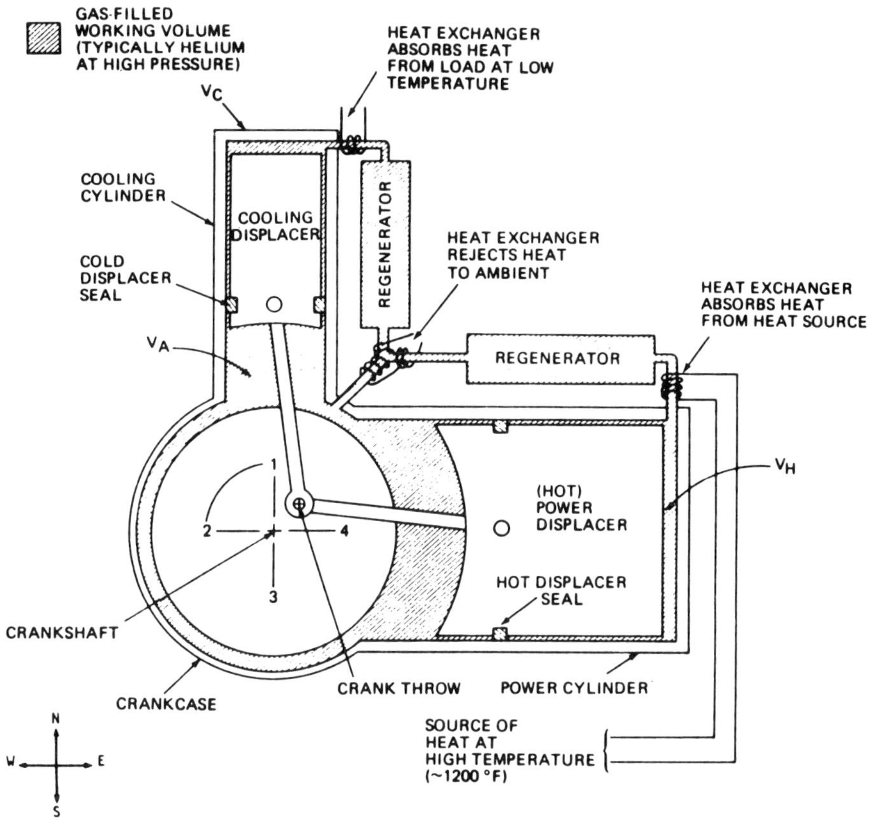

Fig. 4.18 Schematic diagram of a Vuilleumier cryocooler (after White 1976).

The two displacers are coupled through a kinematic linkage to maintain the proper phase relationship. Basically the phase requirement is that the cold displacer is descending as the pressure of the working fluid is decreasing. This corresponds to an expansion process in the cold cylinder with the result that heat is absorbed from the cold cylinder surrounds (the refrigeration effect).

The reciprocating elements are displacers (no work devices) rather than pistons so in theory no work input is required to operate a Vuilleumier unit. In practice, a small shaft power input is required to overcome the small aerodynamic and mechanical friction effects.

The schematic diagram of a Vuilleumier cryocooler shown in Fig. 4.18 indicates the ambient temperature spaces of the hot and cold cylinders below the displacers are common and constitute the 'crankcase' of the

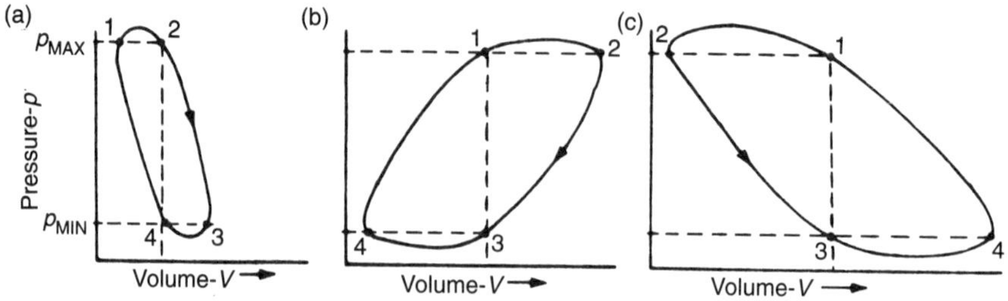

Fig. 4.19 Pressure–volume (or work) diagrams for the (a) cold volume, (b) hot volume, and (c) ambient temperature volume of an idealized Vuilleumier cryocooler (after White 1976).

unit. Heat is rejected at near ambient temperature from the cryocooler in the crankcase area. Heat is absorbed by the hot cylinder (the heat input) at high temperature and is absorbed by the cold cylinder (the refrigerating effect) at low temperature.

Pressure–volume or work diagrams for the three spaces of an idealized Vuilleumier cryocooler (with isothermal heat transfers and no fluid friction effects), as determined by White (1976), are reproduced in Fig. 4.19. The work diagrams for the hot and cold spaces are clockwise in direction indicating positive work by the gas on the displacer whereas the diagram for the ambient temperature space is counterclockwise indicating negative work by the gas on the displacer. The sum of the areas for the hot and cold spaces exactly equals the area for the ambient temperature space, that is, there is no net work input or output. Under the idealized isothermal conditions of operation the area of the spaces also represents the heat flow to the hot space and the cold space as well as the heat flow from the ambient temperature space. The diagrams shown in Fig. 4.19 are perhaps somewhat misleading for they suggest the pressure ratio p_{max}/p_{min} to be very high, perhaps as much as 20, assuming the lowermost point on the pressure scale is a true zero. This is not the case. In fact the pressure ratio of Vuilleumier machines is, characteristically, very low, about 1.05:1 compared with the typical ratio of 2:1 of Stirling cryocoolers.

The basic technology of Vuilleumier cryocoolers has been well presented by White (1976) and summarized by Walker (1983).

4.10.1 Commercial applications of Vuilleumier cryocoolers

Vuilleumier cryocoolers enjoyed a substantial measure of research and development support principally by the US Air Force from the mid-1960s

Fig. 4.20 Split-Vuilleumier cryocooler model no. VM-1, capacity 1 watt 77 K (courtesy Cryogenic Technology Inc.).

through the 1970s. They were attractive in miniature configurations principally because of the promise of high reliability, arising from the low force loadings, as a consequence of the low pressure ratios achievable with a thermal compressor (compared with the mechanical compressor of the Stirling unit). In course of this research/development effort various problems peculiar to the Vuilleumier cooler were encountered particularly those concerned with unlubricated rubbing elements at the hot end. In the main these were eventually overcome but the dramatic superiority for extended operation of Vuilleumier cryocoolers was not realized in practice. They tended to be larger, heavier, require a higher energy input, a larger cooling system and to cost a good deal more to manufacture than the Stirling cryocoolers with which they were competing.

As a consequence, nowadays the bloom is definitely off the Vuilleumier cherry although one machine remains in series production for the US Air Force. This is the split-Vuilleumier cryocooler illustrated in Fig. 4.20 with the characteristics summarized in Table 4.2.

Table 4.2 Specifications for CTI Model No. VM-1 Split-Vuilleumier Cooling Engines

Refrigeration types	Split-Vuilleumier cycle
Refrigeration capacity	10 W @ 77 K (25 °C ambient 0.7 W @ 77 K (55 °C ambient)
Cool-down time to 85 K	⩽10 min (55 °C ambient)
Weight	10.0 lb (4.54 kg)
Power requirements	115 Vac, 3ϕ, 400 Hz, 370 W
Ambient temperature limits	Operating: −51 °C to 55 °C Storage: −54 °C to 71 °C
Scheduled maintenance	2200 operating hours
Military specifications	Meets applicable US Military specifications

Despite their failure to compete effectively in the market-place some remarkable Vuilleumier cryocoolers were successfully demonstrated and tested for long maintenance-free periods. The most impressive, without doubt, was the triple-expansion, high-capacity, Vuilleumier cryocooler illustrated in Fig. 4.21. This was developed by the Hughes Aircraft Corporation, Los Angeles, for the US Air Force Wright Patterson Air Force Base. Further details of this and many other prototypes or limited-production Vuilleumier cryocoolers were given by Walker (1983).

There appears to be little present research and development work in progress on low-capacity Vuilleumier cryocoolers so their commercial future is by no means bright. However, there is considerable research and development interest in relatively large Vuilleumier machines operating at high (near ambient) refrigeration temperatures. These are planned for domestic and commercial natural gas-fired refrigeration, heat pump, and air conditioning applications. In these applications the commercial success of Vuilleumier machines seems assured as:

(1) public concern grows about the use of Freon refrigerants and their impact on the environment; and

(2) natural gas becomes increasingly the fuel of the future in many parts of the world.

Fig. 4.21 Triple-expansion, high-capacity Vuilleumier cryocooler (courtesy Hughes Aircraft Corporation).

4.11 Pulse-tube cryocoolers

4.11.1 Basic pulse-tube refrigerator

In the early 1960s Gifford and Longsworth (1963) discovered a refrigeration technique which permitted elimination of the displacer from the Stirling refrigerator. They called this new technique 'pulse-tube refrigeration'. Under Gifford's direction the pulse-tube refrigerator was advanced to the point where temperatures of 124 K in one stage and 79 K in two stages were achieved. A single-stage unit precooled to 65 K achieved a minimum temperature of 30 K (Longsworth 1967).

A schematic diagram of the basic pulse-tube refrigerator is shown in Fig. 4.22. The version developed by Gifford was closed at the hot end and did not have the orifice and reservoir volume.

The principle of operation, as given by Gifford and co-workers and by Lechner and Ackerman (1973) is qualitatively simple. The pulse tube

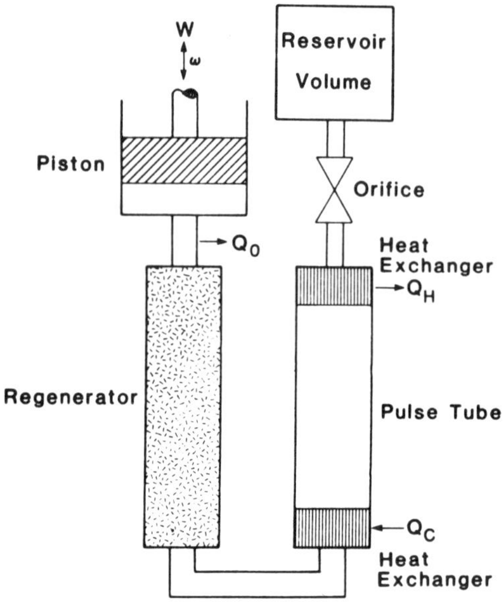

Fig. 4.22 Pulse-tube refrigerator.

is closed at the top end and is equipped with an extended surface to facilitate heat transfer between the working gas (helium is best) and the surroundings in order to dissipate heat Q_H to be rejected from the system at near ambient temperature. The bottom end of the pulse tube is open and also equipped with a good heat transfer surface to absorb the refrigeration power Q_C. The cold open end of the pulse tube is connected to a pressure wave generator via a regenerator. During the compression part of the cycle the elements of gas in the pulse tube move toward the closed end and, at the same time, experience a temperature rise due to the adiabatic compression. At that time the pressure is at its highest value, P_{max}.

During the plateau in the pressure oscillation the gas is cooled somewhat by heat transfer to the tube walls. Then in the expansion part of the cycle the same element of gas moves toward the open end of the pulse tube and experiences a cooling due to the adiabatic expansion.

During the plateau at minimum pressure the cold gas is warmed by heat transfer from the tube walls. The net result of cycling the pressure in this manner is a shuttle heat transfer process in which each element of gas transfers heat toward the closed end of the pulse tube. The heat pumping mechanism described here requires that the thermal contact

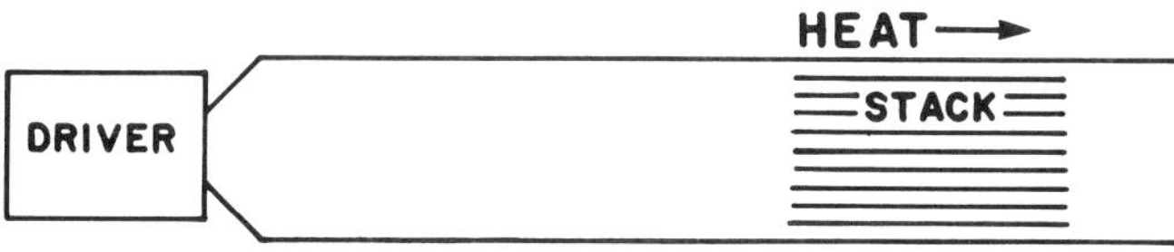

Fig. 4.23 Resonant pulse-tube (thermoacoustic) refrigerator.

between the gas and the tube be imperfect so the compression and expansion processes are somewhere between isothermal and adiabatic. It has been found that the highest intermediate heat transfer occurs generally when the thermal relaxation time τ_t between the gas and the tube walls is such that $\omega\tau_t = 1$, where ω is the angular frequency of operation.

4.11.2 *Resonant pulse-tube (thermoacoustic) refrigerator*

Merkli and Thomann (1975) showed that when a gas-filled tube, closed at one end, is driven at resonance with an oscillating piston at the other end, cooling of the tube walls occurs at positions near the centre of the tube (pressure node) and heating occurs at the ends (pressure antinode). No internal structure was used in the tube and the high frequency of 122 Hz resulted in nearly adiabatic conditions ($\omega\tau_t \gg 1$) so little refrigeration was produced. More recently Wheatley *et al.* (1985) improved the resonant pulse-tube or thermoacoustic refrigerator by adding internal structure at the appropriate position along the tube to get $\omega\tau_t = 1$ (see Fig. 4.23). The internal structure of the 39 mm diameter tube consisted of cloth-epoxy plates 67 μm thick and spaced by 380 μm. A frequency of 516 Hz was used with a mean pressure P_m of 1 MPa and a dynamic pressure amplitude of 0.034 P_m. The lowest temperature achieved with this type of device was 195 K. The Wheatley device has no regenerator and the thermodynamically active region in the gaps between plates is positioned differently from that of the basic pulse tube. However, there is no fundamental reason why the resonant pulse tube cannot use the same geometry, including a regenerator, as the basic pulse tube.

4.11.3 *Orifice pulse-tube refrigerator*

Mikulin *et al.* (1984), a professor at the Moscow Technical High School, the Baumann Institute, added an orifice at the top of the basic pulse tube to allow passage of some of the working fluid through the orifice to a large reservoir volume. Figure 4.22 shows a later version of this modification developed at the US National Bureau of Standards (Radebaugh 1987). The original work of Mikulin *et al.* placed the orifice just below the isothermal section instead of above it as in the Radebaugh version. They were able to achieve a low temperature of 105 K using a pulse tube 10 mm in diameter by 450 mm long with pressures $P_{max} = 0.4$ MPa and $P_{min} = 0.2$ MPa at a frequency of 15 Hz. The net refrigeration capacity at about 120 K was 10 W.

Experiments were done by Radebaugh at the National Bureau of Standards using the modified geometry of Fig. 4.22. The orifice was a needle valve which could easily be adjusted to give the minimum temperature. The pulse tube was a thin-wall stainless steel tube 12.7 mm outside diameter by 237 mm long. A 22 mm long copper section was soldered to the top end and packed with 80 mesh copper screen to enhance heat transfer to the water-cooling coil wrapped on the outside of the copper section. The reservoir volume, large when compared with the pulse tube, was 1 litre.

Optimum orifice adjustment gave a low temperature of 60 K with $P_{max} = 1.24$ MPa and $P_{min} = 0.71$ MPa ($P_{max}/P_{min} = 1.75$) and a frequency of 9 Hz. Gross heat pumping values from the cold to hot ends were about 18 W at 60 K with those conditions.

4.11.4 Multiple-stage pulse-tube cryocoolers

It is easy to add any number of stages, still using one pressure wave generator, for the basic and orifice types of pulse tubes. The hot end of the second-stage pulse tube is attached to the cold end of the first stage. A second regenerator is added below the first with some gas passing through it to the second stage.

Early experiments on multiple stages was reported by Longsworth (1985). He obtained 79 K with two stages of a basic pulse tube and 32 K with four stages. Radebaugh estimates it may be possible to reach 10 K with four stages of the orifice pulse tube provided sufficiently good regenerators can be incorporated.

4.11.5 Comparison with other refrigerators

Radebaugh (1987) compared the intrinsic cooling efficiency of orifice pulse tubes with those of the Stirling and Joule–Thomson refrigerators. His findings are shown in Fig. 4.24. The pulse tube efficiency was slightly less than the Stirling refrigerator because the reversible motion of the displacer has been replaced with irreversible expansion.

The refrigeration power per unit mass flow is about three times less than an ideal Stirling refrigerator. The smaller capacity for the pulse

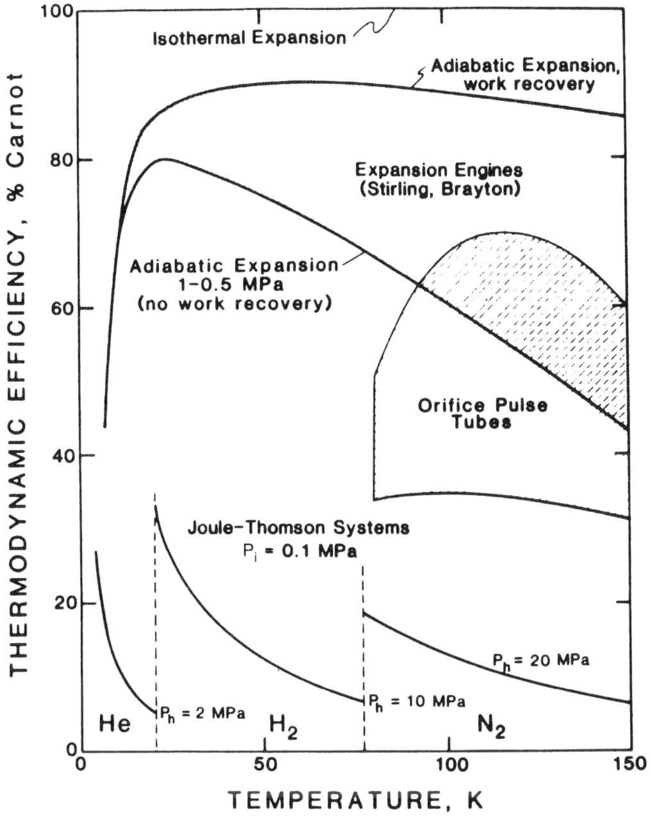

Fig. 4.24 Comparison of intrinsic cooling efficiency for various refrigerators (after Radebaugh 1987).

tube means that, for the same refrigeration as in a Stirling refrigerator, much larger mass flows through the regenerator must occur. The resulting larger regenerator loss term degrades the overall system efficiency unless improved regenerators are used. However, the Stirling refrigerator has shuttle and frictional losses associated with the displacer motion that are not present in a pulse tube. A further advantage of the pulse tube is that any vibration of the cryocooler associated with displacer motion in the Stirling engine is eliminated in the pulse tube.

In future developments the operating principle of the basic pulse tube can perhaps be extended to other systems. One example is the magnetic system briefly mentioned by Radebaugh (1987) where a rod of magnetic material oscillates in a magnetic field gradient to produce refrigeration at the low-field end.

4.11.6 Commercial application of pulse-tube cryocoolers

At the present time (1993) there are no established commercial applications for pulse-tube cryocoolers, indeed there appears to be no pulse-tube refrigerators of any description in series production anywhere. There is, however, considerable intense interest in the potential of pulse-tube refrigerators for long-endurance, low-maintenance applications. This appears more achievable with pulse-tube refrigerators than with their principal competitor, Stirling refrigerators, because of the simplification effected by elimination of the displacer from the Stirling engine to convert it to a pulse tube. An increase in the size and weight and a decrease in efficiency may be an acceptable price to pay for a significant extension of maintenance-free operation.

Recent work by Radebaugh and Swift of the Los Alamos National Laboratory combined a thermal acoustic driver (to produce the pressure pulsations) and an orifice pulse-tube refrigerator (christened TADOPTR) to achieve a cryocooler *with absolutely no moving parts*. Unfortunately the pressure generator thermal acoustic device (TAD) required a tube about 5 cms diameter wound in a coil about 3 m diameter so the device appears far from practical where space, size, weight, and, probably, cost are important parameters.

Six papers on various aspects of pulse-tube cryocoolers were presented at the Seventh International Cryocooler Conference in 1992 including an analytical and experimental study of a multistage machine capable of achieving 4 K (Matsubara *et al.* 1993).

4.12 Comparison of the cycles: similarities and differences

Five different diagrammatic arrangements for mechanical cryocoolers are shown in Fig. 4.25. The principal common feature is that they all include a regenerative heat exchanger, a porous matrix of finely divided material with a temperature gradient ranging from ambient at one end to the minimum cycle temperature at the other.

Another common feature is they all can be used for cryogenic refrigeration. All can achieve temperatures in the cryogenic range with a single stage of expansion, commonly 80 K or better, and lower temperatures with several stages of expansion, 20 K with two stages, 6 K with three or four stages. They all use a working fluid which remains gaseous throughout. There is no change of phase from gas/vapour to liquid. Generally, and especially for the lower temperatures, the only available working fluids are helium or hydrogen. Hydrogen is exceptionally difficult to contain and is, moreover, very highly reactive so is used only rarely.

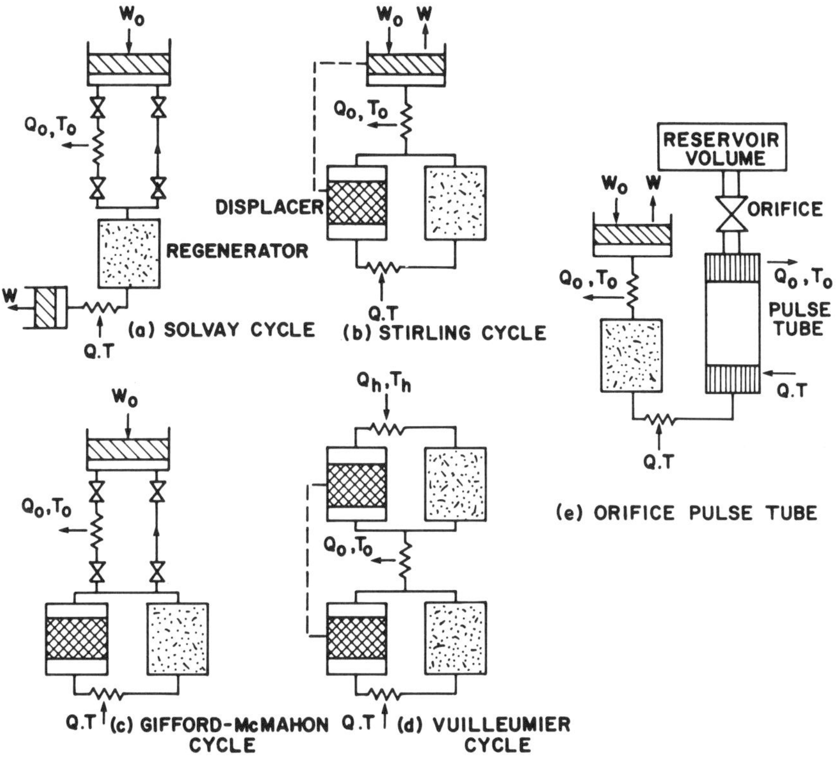

Fig. 4.25 Composite diagram of five different arrangements of mechanical cryocoolers all incorporating a regenerative heat exchanger.

All the arrangements incorporate recuperative heat exchangers to abstract heat at low temperatures (the refrigeration process) from the surrounds of the cold expansion cylinder to the working fluid and to reject heat at near ambient temperature (the cooling process) from the working fluid to the coolant, air or water.

Further, all the arrangements incorporate reciprocating pistons or displacers. Pistons are heavy-duty structural elements with a large pressure difference across them transferring work to and from the working fluid and the kinetic linkage. They generally operate at near constant temperature and may be described as 'high Δp, low ΔT devices'. Displacers are light structural elements that do no work, but simply displace gas from one space or another, usually at different temperature levels. Displacers may be described as 'high ΔT, low Δp devices'. It is not uncommon, particularly on small low-capacity units, for the regenerator to be incorporated in the displacer.

All arrangements incorporate either two pistons, two displacers, or one of each, except the pulse tube which uses only one reciprocating element, a piston operating at near ambient temperature, a singular advantage of the pulse tube.

Four of the five arrangements use mechanical compressors to generate the pressure fluctuations in the gas and so require an input of mechanical work (almost without exception provided by an electrically driven motor). The Vuilleumier arrangement operates with a thermal input at high temperature and requires only a modest input of work to maintain phasing of the two displacers. Frequently on small cryocoolers the heat input is provided by electrical resistance heating. This of course is a thermodynamic abomination for the Carnot penalty (an effective loss of 60 per cent of the input energy) must be paid to generate the electricity. However, for low-capacity cryocoolers the convenience of electric resistance heating is worth the price it exacts.

Two arrangements, the Gifford–McMahon and the Solvay, incorporate valves to regulate and control the flow of working fluid to and from the cold expansion part of the machine. The use of valves is at once a blessing and a curse. Valves are a complication and an additional source of trouble that give valveless machines a definite advantage, particularly in the miniature arrangements used for infrared systems and other cold electronics applications at near liquid nitrogen temperatures.

Nevertheless, the use of valves does offer the powerful advantages of independent selection of the maximum and minimum cycle pressures. In valveless machines the *mean* pressure depends on the charge pressure of gas admitted and can be independently chosen. However, the *pressure ratio* and the way the fluid moves internally depends on the internal geometry of the unit. It is rarely possible to devise an arrangement for a valveless machine that gives a pressure ratio, p_{max}/p_{min}, greater than 2:1 for a valveless cryocooler. This limiting value does not apply to the valved Gifford–McMahon and Solvay machines. Any pressure regime desired may be selected by the designer.

Moreover, the use of valves permits the physical separation, to a virtually unlimited extent, of the smaller, lightweight, cold expansion parts of the cryocooler from the heat, noise, and mechanical vibration of the compression parts. Similar split systems can be devised for the valveless Stirling or Vuilleumier arrangements with the advantage that only a single transfer line is necessary. However, in this case the cooling performance is significantly and progressively affected by fluid losses in the transfer line to the extent that the length and diameter of the transfer line are in practice severely limited to a maximum of about 2 metres.

With a valved machine the speed of operation and the type of compressor and expander (reciprocating or rotary) can be independently selected. It is not uncommon to find, in Gifford–McMahon cryocoolers, a relatively high-speed rotary compressor used in conjunction with a very slow reciprocating expander. This is attractive as the way to reduce the size and weight of the compressor with greater reliability and lower cost while retaining the advantages of a slow-speed, long-life, vibration-free cold end.

In valveless machines the compressor and expander are both, almost without exception, reciprocating machines operating at the same frequency. While advantageous in small low-capacity units it is a grave handicap that becomes increasingly severe with larger machines of higher refrigeration capacities.

The consequence of these differences plus the interplay of commercial and military pressures is that Stirling engines have come to dominate fields where low-capacity (0.1 to 1 watt at 80 K) compact, efficient, low-cost machines are required. These are primarily the military and civil applications of infrared night vision and missile guidance systems. The penalty for possession of these attractive features is the limited operating life of miniature Stirling systems, commonly less than 500 hours but pushing upwards to 3000 hours with the new, close-tolerance seal, linear motor, load demand models now entering production.

Gifford–McMahon and modified Solvay cryocoolers have come to dominate in those fields of application where long-term reliability (major maintenance intervals of 10,000 hours) is the most important criterion. They are particularly used in double expansion versions capable of achieving 20 K refrigeration for cooling applications such as cryopumping or the precoolers for helium liquefiers in conjunction with a Joule–Thomson expansion stage.

The price to be paid for these advantages is a big, heavy unit of relatively high capital cost and high power input. In terms of units manufactured the number of Gifford–McMahon machines is a small fraction of the thousands of miniature Stirling cryocoolers made every week in the United States, Europe, Israel, and, presumably, the Eastern bloc countries.

The above comparison drawn of the two basic types of regenerative cryocooler, one with valves and one without valves and the implication of their consequent gravitation to their natural fields of application is probably over simplistic. There is no intrinsically inherent factor in the division of application that has come about. It has depended much on the diligence, innovation, and commercial acumen with which the various manufacturers have pursued the development and marketing of their individual designs.

Having made the distinctions we have above one may then readily find exceptions to the rule. For example, large-capacity versions of Stirling cryocoolers (1 kW and 4 kW at 80 K) are manufactured, and have been since the early 1950s, as well as multiple-expansion units achieving lower temperatures than 80 K Similarly, compact Gifford–McMahon units of surprisingly small size have been made in research/development programs although, so far as is known, none have passed into series production.

4.13 Materials

A variety of materials are used in the fabrication of mechanical cryocoolers. Stainless steel is the most widely used material. Stainless steel has many advantages for cryogenic use. It is an alloy of iron with substantial proportions of chromium and nickel, typically 18 and 8 percent, respectively. It is strong, ductile, with a low coefficient of thermal conductivity and is readily available in a wide variety of shapes and sizes at modest cost. It is uniformly homogeneous and high quality with well-established specifications and performance. Stainless steel can be machined and shaped with standard cutting tools and procedures and can be joined relatively easily.

An important advantage, moreover, is that stainless steel is not subject to ductile–brittle transformation at low temperatures. This transformation occurs with many ferrous materials making them unsuitable for use at low temperatures.

In ductile–brittle transformation the material when cooled to a critical temperature characteristic of the material loses the ability to absorb even low levels of shock and impact loadings which at higher temperatures would have no effect on the material. At low temperatures modest impact loadings cause the material to break like a carrot exhibiting brittle fracture rather than the normal ductile behaviour.

Stainless steel is non-magnetic, an important property in some applications and has a low thermal conductivity. Where higher values are required copper and aluminum are widely used instead of stainless steel. Copper is heavy but compared with aluminum very easy to join using solder joints.

The principal non-metallic material used in mechanical cryocoolers is the fibreglass-reinforced, epoxy–plastic material, 'G10'. Other plastic materials include nylon and, principally because of its low friction characteristics, polytetrafluoroethylene, PTFE or Teflon,† often reinforced

† Teflon—trademark of Du Pont de Nemours, Delaware.

with a filler such as chopped glass, carbon, powder metal, or a plastic amide material in the well-known series of 'Rulon'† plastics widely used for seals, bearings, gaskets, and so on.

4.14 Historical perspective

4.14.1 Stirling cryocoolers

The early days The first engine to incorporate use of a regenerative heat exchanger was patented by the Revd. Robert Stirling in 1817 for application as a power system. The same cycle for use as a refrigerator was proposed by John Herschel as early as 1834 and successful refrigerators operating on the reversed Stirling cycle were subsequently described by Alexander Kirk (1874). They were said to have been in use at Bathgate, Scotland, for over 10 years.

Stirling power systems, described in those early days as *hot-air engines* enjoyed considerable commercial success throughout the nineteenth Century. They were safer and easier to use than contemporary reciprocating steam engines. Both were displaced about the turn of the century by the invention of internal combustion engines and the electric motor. The use of Stirling engines as refrigerators was, so far as is known, never commercialized beyond the Kirk machine mentioned above.

The Philips program By the time of the First World War Stirling engines had more or less passed out of common experience. Then in the late 1930s, the Philips Electrical Company, Eindhoven, Holland, began the development of a small kerosene-powered generator to recharge the batteries of domestic radio sets. In those days radios were a formidable array of vacuum tubes and other sizeable components with a hefty power drain.

For the power generator Philips selected the hot-air engine for development. In this work they were completely successful and within five years had achieved an order of magnitude increase in power density, speed, and efficiency compared with the earlier hot-air engines. This was accomplished by the application of modern materials and techniques. However, by the time the engine was ready for production, transistors had been invented and battery-powered radio sets capable of operating for long periods without recharging were available. For this reason, the small generator set was never made beyond an initial preproduction run of two hundred units. Cognizant of the singular advantages of Stirling

† Rulon—trademark of the Dixon Corporation, Rhode Island.

engines Philips continued their development of Stirling power systems with engines of larger size and higher power for marine, automotive, and stationary power. They continued this progress until the late 1970s and the work continues today in other parts of the world (Walker *et al.* 1993).

Philips air liquefier The Philips Stirling air liquefier shown in Fig. 4.9 came about as a by-product of the Philips power systems work. When Philips' researchers found their early engines worked well as refrigerators a separate team to develop the refrigerator was established. Work started about 1948 and, under the brilliant leadership of Dr Jan Kohler, the 1 kW (80 K) liquefier was developed and introduced to commercial service about 1954.

There was no apparent target for this unit. It was a completely new device, the like of which had never been seen before. It offered the possibility for anyone with a relatively modest sum of money and effort to produce liquid air at will with nothing more than an input of electrical power and cooling water.

Under Kohler's direction many other auxiliaries and devices based on this unit were developed for commercial service. These included fractionating columns to produce liquid nitrogen from the air liquified by the cooler, cold boxes for cryogenic freeze processing and preservation, multicylinder versions producing kilowatts of cooling, cold transfer systems, multiple expansion machines with a two-stage refrigerator capable of achieving 12 K, and a helium liquefier.

Many of these developments were initiated by Kohler. He contributed substantially to all of them and also made important contributions to the theory of the Stirling cycle. He is justly called the father of the Stirling cryocooler.

Such developments would not be possible in today's world where companies, led by accountants, need to be satisfied by 'market surveys' of an assured profit. At the time the Philips Stirling cryocooler was developed there was no market, for no one knew or imagined the possibility of such a device.

Production of the Philips liquefier commenced in Eindhoven in the early 1950s and has continued there ever since. The number of liquefiers made of the type shown in Fig. 4.9 is unknown but various estimates put the number as high as 15,000 over the past 40 years. They may be found all over the world. They are presently made in Eindhoven and marketed in the United States by International Process Systems and by the Philips organization throughout the rest of the world.

Miniature Stirling cryocoolers The possibilities inherent in the Stirling cryocooler for infrared service soon triggered off a rash of development programs for miniature cryocoolers largely sponsored by the US Army, Navy, Air Force, and the NASA. These developments occurred simultaneously with competitive programs on miniature Joule–Thomson, Vuilleumier, Gifford–McMahon, and Claude cycle machines.

A number of interesting machines resulted from these programs with the development of progressively smaller, lighter, and more efficient miniature refrigerators at a half-dozen different manufacturers and developers. Miniature Stirling cryocoolers have emerged as the clear winners among the miniature mechanical systems required for cryogenic sensors and cold electronics.

Split-Stirling cryocoolers The split-Stirling concept was conceived in the late 1960s, apparently simultaneously by Walt Higa of the Jet Propulsion Laboratory, Pasadena, California, by Fred Chellis of Cryogenic Technology Inc., Waltham, Massachusetts, and by Stuart Horn of the US Army Night Vision Laboratory, Fort Belvoir, Virgina. This invention offered the possibility to separate the heavy compressor and drive motor from the cold, expansion part of the engine. This was particularly attractive for gimbal-mounted infrared sights to reduce the inertia of the system and hence the forces necessary for rapid accelerations and decelerations.

Like many recent developments in Stirling engines the split-Stirling was the reinvention of an earlier idea by Ossian Ringbom, who in 1912 conceived a Stirling power system with a free displacer not connected mechanically to the engine crank mechanism. Ringbom–Stirling engines are currently a topic of interest in the Stirling power system field. Senft (1992) has recently published the first definitive monographs devoted to Ringom or split-Stirling machines summarizing his substantial theoretical and experimental contributions to the technology over the past decade.

Brushless DC motors The development of electronic systems and techniques since the Second World War resulted in a dramatic reduction in the size of infrared detectors and, consequently, in the cooling capacity required. As a result, despite the development of progressively smaller versions, the cryocoolers have remained grossly mismatched to the detectors, offering cooling capacities generally far in excess of that required.

One beneficial consequence of electronic system improvements has been the development of brushless DC motors. For these units a direct current (DC) input from a battery or power supply is 'chopped' into a high-frequency alternating current (AC) input to the drive motor allowing the use of a lightweight brushless AC motor instead of the conventional DC motors with carbon brushes and a commutator system.

Early in the development of Stirling cryocoolers the drive motors were incorporated in the crankcase to eliminate the seal problem where the crankshaft emerged from the crankcase. Unfortunately, in the super-dry helium atmosphere of the crankcase the carbon brushes experience rapid wear contributing much dust and wear debris to the internals of the system. This eventually accumulates in the regenerator blocking the interstices of the matrix causing a severe decline in cooling performance. The dust particles also cause problems in the operation of gas seals, and other close-tolerance areas.

Elimination of the carbon brushes resolved this problem very satisfactorily. Condensation of the solvent vapours emanating from the insulation varnish of electrical coils is unresolved but much can be accomplished by extended 'bake-out' of the unit at elevated temperatures and high vacuum conditions.

The common-module concept In the early days of their development in the United States small coolers were generally developed to suit the required application. The demand was rarely sufficient to justify series production of any particular cooler nor were the funds or time usually available to permit the several generations of developmental prototypes normally required to realize a completely successful unit. Eventually the need for basic designs of cryocooler was recognized by the various military sponsors particularly the US Army and in the 1970s the 'common module' concept emerged. This has found expression in the series production of four basic Stirling coolers, two of the integral and two of the split variety, having nominal capacities of $\frac{1}{4}$ watt and 1 watt at 80 K.

In all these machines the available cooling capacity of the cryocooler is the small fraction of the total or gross refrigerating capacity not consumed by the various parasitic losses in operating the machine to reduce the tip of cold finger to cryogenic temperatures. The available refrigeration is then the very small difference of the gross refrigeration and the combined losses. The resulting consequence is that ostensibly identical machines have significant differences in the minimum temperature they can achieve ranging in units having a single stage of expansion from, at best, a low of 40 K to, at worst, a high of 140 K.

Demand control Until the late 1980s, most Stirling cryocoolers had no control mechanism whatsoever. They were simply switched on or off. The refrigeration capacity required to cool the cold finger to its operating temperature greatly exceeds the capacity required to maintain it at the desired temperature. Thus a cryocooler equipped with sufficient capacity to accomplish cool-down in a reasonable time following start-up generally produces far more cooling than is necessary to maintain the cold finger temperature with consequent excess power consumption. To overcome this problem contemporary developments have equipped coolers with a miniature electronic load demand system sensing the temperature of the cold finger or infrared sensor and regulating the speed of the drive motor to provide just sufficient refrigeration to maintain the temperature at the set point. As a consequence the 'cruise' power input is reduced, typically to only 15 watts, compared with the 30 watt power input during cool-down of a nominal $\frac{1}{4}$ watt, 80 K cryocooler. The reduced speed of operation also extends the operating life of the mechanical parts, bearings, seals, and so on.

Integrated detector/cryocooler Until recently the cryocoolers and the infrared detectors in their associated dewar flask encapsulation were made by different specialist manufacturers. Often the union of the two assemblies was consummated at the premises of a systems manufacturer or in field replacement of one or the other unit. This often resulted in a difficult 'interface problem' with the cold finger of the cryocooler poorly fitted to the dewar and resulting frequently in damage to the expensive and delicate glass dewar or, at best, a deterioration of detector performance because of improper cooling.

This problem has now been overcome by the development of integrated assemblies wherein the infrared detectors are applied directly to the end of the cooler cold finger and the dewar encapsulation is applied to the completed, sealed, cooler.

Hermetic sealing Helium is a highly mobile gas, particularly under pressure. It will over time find a way to escape through virtually any type of static seal. To improve shelf storage life and to avoid also the possibility of unscheduled, well-meaning but often unskilled field maintenance there is a tendency for all joints to be welded shut. This effectively seals the unit for life and ensures that repairs can only be made at a facility properly equipped to do so. In most units provision is made for the working fluid (helium) to be replaced at specified intervals. Contamination of the working fluid, leading to a progressive loss of performance over time, occurs because of the release of internal vapours, particularly water from the constituent materials, and the generation of carbon dioxide in the normal operation of bearings and other metal contacting surfaces due to the decomposition of lubricants.

Close-tolerance seals In the past the operating life of Stirling cryocoolers has been very limited, ranging from a few hours to as much as a thousand hours. The principal cause of this limited life was the rubbing seals on the reciprocating elements, both in the cold space and the hot space.

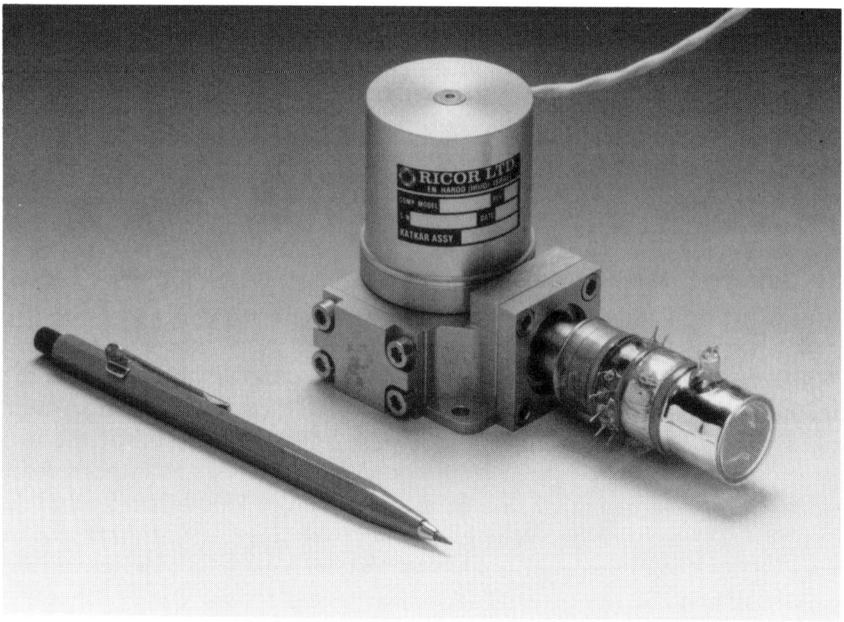

Fig. 4.26 Superminiaturized Stirling cryocooler (courtesy Ricor Inc., Israel).

Most of the plastic dry rubbing materials operate best with a trace of water vapour in the environment. Unfortunately any water vapor present precipitates in the cold regions of the cryocooler. Consequently the very dry helium atmospheres encountered in miniature Stirling cryocoolers are a particularly demanding environment for rubbing seals.

The life-limiting dry rubbing seals have now been largely replaced by close-tolerance seals with consequent significant increase to several thousand hours in the operating life.

Linear electric drivers Adoption of close-tolerance seals has required elimination of the piston-side forces encountered in conventional crank-driven compressors with rotary electric drivers. This has been accomplished by the replacement of rotary electric drivers with linear electric motors. However, at this time, linear electric motors are not so highly developed as rotary electric motors. They are bigger, heavier, less efficient, and more difficult to balance dynamically. Moreover, in limited production they are more costly than the rotary, brushless DC motors commonly used. Impressive strides in the resolution of these problems has been made as the move to close-tolerance seals and linear electric drivers has accelerated with more manufacturers in the field and the incorporation of improvements resulting from intense research and development efforts.

Superminiaturized Stirling cryocoolers As infrared detectors become smaller and smaller the cryogenic cooling requirement has decreased concomitantly. To meet the demand for reduced capacity a new generation of superminiaturized Stirling cryocooler has emerged.

At the forefront of these developments is the Ricor cryocooler shown in Fig. 4.26. This novel machine, an integrated two-piston vee Stirling arrangement was designed, developed, and manufactured by Ricor Ltd. of the En Harod Kibbutz in Israel.

Theoretical analysis and computer simulation The analysis of Stirling cryocoolers has benefited much from the intense efforts invested to develop theoretical techniques for analysis of Stirling power systems. Much of this work can be immediately carried over to Stirling refrigerators.

A summary of the status of theoretical developments including computer simulation was given recently by Walker (1989). This includes an account of a Stirling cryocooler design aid, CRYOWEISS, developed at the University of Calgary for use on microcomputers by non-computer specialists. The program CRYOWEISS is available at modest cost.† It is carefully written to be easy to use and includes an optional novel graphics package which greatly facilitates the rapid optimization of a given design.

Future developments The anticipated development of cold electronics systems, both semiconductor and superconducting, requiring miniscule cooling at 80 K presages a successful commercial future for superminiaturized Stirling cryocoolers. The development of larger-capacity Stirling cryocoolers of 0.5 to 1 kW (at 80 K) capacity and multiple-expansion versions capable of lower temperatures including helium liquefaction is occurring at half-a-dozen locations in the US, Europe, and Japan. Various models are expected to become commercially available in the early 1990s.

One application for these units is to reliquefy the nitrogen vapour generated by supercomputers operating with their very large scale integrated circuit boards operating at cryogenic temperatures submerged in liquid nitrogen. Another growing application is the requirement for reliquefaction of helium vapour or low-temperature cooling to reduce

†$500 US with graphics package, from Marvin Weiss, Department of Mechanical Engineering, University of Calgary, Alberta, Canada or Brad Ross, Stirling Machine World, 1873 Hummingbird Court, Richland, Washington.

the helium boil-off in a variety of medical and scientific equipment now going into service.

Although they were the first in the field with Stirling cryocoolers of the type shown in Fig. 4.9 Philips did not maintain the impetus of their original efforts. They appear not to have carried out the research and development necessary to endow these machines with the capability to operate for a year or two unattended and requiring no significant maintenance attention. Instead they continued the production of original designs with little further development so the maintenance-free life is limited to 1000 hours or less. They did however do substantial research in linear cryocoolers of low capacity. Through their United States affiliate, North American Philips Laboratories, they were at the forefront of development in miniature Stirling cryocoolers and long-lived space cryocoolers until they retired from the field about 1990. It is unfortunate that their expertise in long-lived miniature Stirling cryocoolers was not carried across to the larger machines of 1 kW capacity or more.

4.14.2 Gifford–McMahon cryocoolers

Gifford–McMahon cryocoolers were invented in the mid-1950s by William Gifford and Howard McMahon while working at A. D. Little Inc., Cambridge, Massachusetts. Two similar machines were invented apparently simultaneously. One involving the production of expansion work was basically the reinvention of the original concept by Solvay. The other, involving the use of a cold displacer and without the production of expansion work, was basically the reinvention of the original concept by Davy Postle. The Postle type rapidly proceeded to commercial utilization. The Solvay type did not pass beyond the prototype demonstration stage.

Chellis (1978) observed that one of the earliest closed-cycle cryocoolers, developed about 1958, to cool a three-channel ruby laser in a target-tracking radar set, was a three-stage expansion Gifford–McMahon cooler acting as the precooler for a low-temperature Joule–Thomson expansion stage providing cooling at 3.8 K. Similar installations were provided to refrigerate the ground-based parametric amplifiers for the Telstar antenna at Andover, Maine, for intercontinental satellite communications and for the NASA spacecraft communication network at Goldstone, California, at Woomera, Australia, and in Madrid, Spain.

Subsequent developments in amplifier performance led to an amplifier capable of operating satisfactorily at 20 K. This relaxed the cooling requirement immeasurably. It permitted the use of a Gifford–McMahon cryocooler with only two stages of expansion and led to the evolution of the present dominant position they enjoy in the market for long endurance and low capacity (a few watts) at 10 to 20 K. Cryogenic Technology Inc. (CTI) of Waltham, Massachusetts, a spin-off company from A. D. Little, has made over 1000 of these units for the satellite communication network. Gifford–McMahon coolers are also made by, among others, (a) Cryomech of Jamesville, New York, a company founded by William Gifford on leaving A. D. Little to take up a professorship at the University of Syracuse, New York, and (b) the A. P. Division of Intermagnetics Inc. of Allentown, Pennsylvania, until recently the Cryogenics Division of Air Products and Chemicals.

Another major market for Gifford–McMahon cryocoolers which developed in the early 1970s was cryopumping, the use of 10–20 K refrigeration to achieve a high vacuum. A substantial increase in this market occurred with the development of semiconductor production with stringent requirements for cleanliness and high-quality processing.

The cryopumping application brought other manufacturers to the field including Balzers High Vacuum, Varian, Inc., Sargent Welch Inc. in the United States, L'Air Liquide in France, Leybold–Heraeus in Germany, and Osaka Oxygen Industries, Suzuki Shokan Ltd., Ulvac Cryogenics Inc., and Toshiba Corporation in Japan. Other Gifford–McMahon cryocooler manufacturers include Cryosystems Inc. in the United States and Cryogenic Consultants Ltd. in England.

Most of the above companies are known to be, or thought to be, actively involved in or, at least, contemplating the opportunities for expansion presented by the requirements for recondensation of helium in the expanding medical, scientific, industrial, and military applications of superconducting magnets with a liquid helium bath. Many are also interested in the possibilities for cold electronics applications, particularly supercomputers requiring substantial refrigeration capacities of up to 1 kW at 80 K.

Endurance problems were encountered with the modified air conditioning compressors adapted for helium service used in early Gifford–McMahon machines. These problems now appear to have been resolved with the development of specialist compressors to serve the expanded market. Maintenance-free intervals of 10,000 hours are said by the manufacturers to be routinely achieved. Information about the experience of users is harder to come by since there appears to be no organized method for collecting and disseminating such information. An attempt by Professor G. Walker to establish a Cryocooler User Experience Center at the University of Calgary evoked an almost total lack of response. Perhaps this means there are (a) no cryocooler users or (b) they have no worthwhile experiences to report. It is, more than likely, that notices of the Center, included in announcements of the 1986 International Cryocooler Conference and the 1987 Cryogenic Engineering Conference, did not come to the attention of those who really knew the information sought: the technicians and technologists charged with responsibility for cryocooler maintenance.

Vuilleumier cryocoolers The first known description of a closed, regenerative, *thermally activated* refrigerating device was given in the United States patent issued in 1918 to Rudolph Vuilleumier, described therein as a citizen of the United States, resident in New York. No other details are known of him or his machine.

The same unit, more or less, was invented independently 20 years later by Vannevar Bush (1938), a professor at the Massachusetts Institute of Technology. Later on K. W. Taconis (1951) of the University of Leiden, Holland, described in a United States patent yet another variant of the heat-powered refrigerator with regenerative heat exchangers. Chellis and Hogan (1964) described a heat-powered, Vuilleumier cryocooler for cooling in the 10–20 K range. It was precooled to 80 K using liquid nitrogen and rejected its heat at this low 'pseudo-ambient' temperature.

The most significant effort to develop Vuilleumier cryocoolers in the low-capacity range for infrared and other cryogenic sensors was invested in the 1970s by Hughes Aircraft Company of Los Angeles with substantial funding provided by the US Air Force, Wright Patterson Air Force Base, Dayton, Ohio. Other companies, notably Cryogenic Technology Inc., Waltham, Massachusetts, and Airesearch Manufacturing Company, Torrance, California, were also involved in Vuilleumier Cryocooler development and production. Other sponsoring agencies included the US Army Night Vision Laboratory, Fort Belvoir, Virginia and the NASA Goddard Space Flight Center (Walker 1983).

It was a valiant effort but the net result of it all is that, at the present time, only one Vuilleumier cryocooler remains in limited production and the future of that is not at all bright. These units are used in the B52 bombers, now an elderly and obsolete force in process of phasing out. Vuilleumier cryocoolers turned out to be larger, heavier, less efficient, and costlier to manufacture than Stirling machines. They do not have the established record of long-term endurance of the Gifford–McMahon machines to compensate. The only advantage they have compared with other cryocoolers is the ability to use a thermal rather than mechanical (electric motor) input. This advantage was rarely utilized in the largely military sponsored programs reported above. They nearly all used electric resistance heating on grounds of convenience.

As a heat pump the Vuilleumier machine does appear to have a brighter future. The heat pump can be used with natural gas combustion thermal input for refrigeration and air conditioning in many domestic and commercial applications. It can operate effectively with air as the working fluid. It thus offers both a means to escape from the use of Freon organic working fluids and has many thermodynamic, space, and weight advantages over conventional thermally heated, absorption refrigerators.

Pulse-tube cryocoolers The pulse-tube cryocooler was invented by William Gifford and his co-workers at the University of Syracuse in the early 1960s. In its first embodiment the closed-cycle pulse tube was equipped with valves to regulate the gas flow to and from the regenerator and the compressor. It then resembled a Gifford–McMahon cryocooler without the displacer.

The system was then simplified by elimination of the valves. This valveless embodiment required the compressor to be mounted close to the regenerator and cold region. The unit then resembled a Stirling cryocooler without a displacer or piston in the cold region.

The minimum temperatures achieved by Gifford and his co-workers was 124 K with a single-stage unit, 79 K with two stages of expansion, and 30 K with a single-stage unit precooled to 65 K Pulse tubes of this type were never developed for commercial production.

Later, in the late 1970s and early 1980s, at the Los Alamos National Laboratory, John Wheatley, Albert Migliori, and Greg Swift investigated thermoacoustic engines in an endeavour to develop a cryocooler with no moving parts. This work has been succinctly summarized by Swift (1988) showing the similarities and differences to the Gifford type pulse-tube.

In 1984 Evgeny Mikulin and co-workers at the Baumann Technical High School and Research Institute in Moscow reported adding an orifice and orifice chamber to the Gifford type valveless pulse-tube. They accomplished a dramatic improvement in performance achieving a low temperature of 105 K compared with Gifford's lowest temperature of 124 K with a single stage of expansion.

This breakthrough rearoused the interest of the cryogenic community in pulse-tube cryocoolers. Following the Mikulin development, Radebaugh and his co-workers at the National Engineering Laboratory (now the National Institute of Science and Technology (NIST)), Boulder, Colorado, have carried out extensive research and development studies on pulse-tube cryocoolers achieving temperatures as low as 60 K with a single stage and less than 30 K with two stages.

Radebaugh (1987) classifies the pulse-tube cryocooler as a 'variation of the Stirling refrigerator. ... The elimination of the moving mechanical displacer has the potential for improved reliability. ... Because of its relatively high refrigeration power, high efficiency, low operating temperature, and simplicity the orifice pulse-tube refrigerator has considerable potential for an alternate cryocooler in the range from 10 to 100 K.'

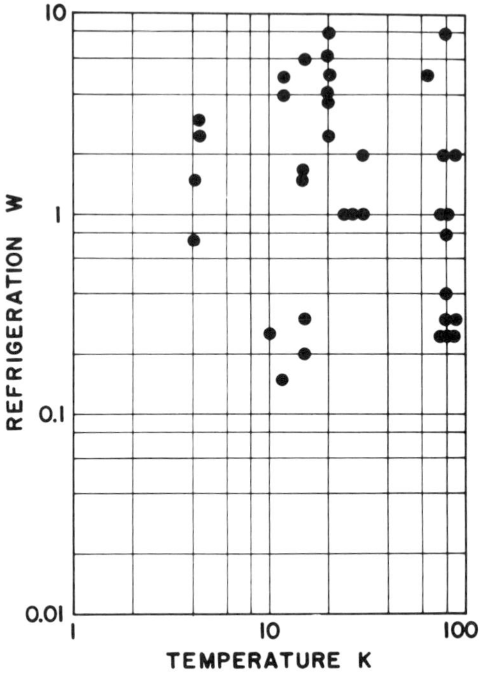

Fig. 4.27 Refrigeration capacity as a function of operating temperature for commercially available miniature cryocoolers (after Smith *et al.* 1984).

At the present time pulse-tube cryocoolers are not in commercial production. However, the present development of 'cold electronics' systems, both superconducting and semiconducting, provides the impetus for superminiaturized cryocoolers of high reliability so the continued development of pulse-tube refrigerators and their early commercialization is a virtual certainty.

Comparative surveys From time to time comparative surveys of the many different cryocoolers are carried out. An early survey of note was carried out by Daunt and Goree (1969) for the United States Office of Naval Research. They compared miniature cryogenic refrigerators and assembled a large compendium of data for the cooling engines then available.

More recently Joseph Smith and co-workers at the Massachusetts Institute of Technology carried out a survey of miniature cryocoolers for superconductive devices (Smith *et al.* 1984). With remarkable prescience considering that, at the time, the discovery of high-temperature

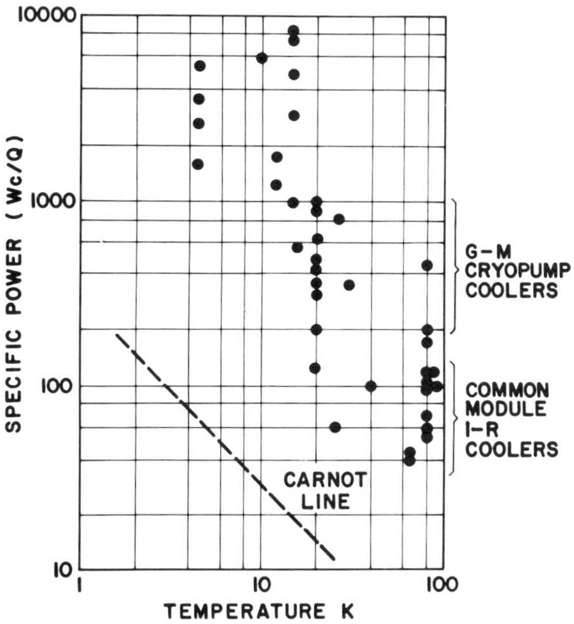

Fig. 4.28 Specific power (work input (watts) per watt of refrigeration) as a function of operating temperature for commercially available miniature cryocoolers (after Smith *et al.* 1984).

(i.e. 80 K and above) superconductors was still two years way and entirely unsuspected, they organized their collection of data to include miniature cryocoolers capable of refrigeration in the temperature range 4–80 K. For their survey Smith *et al.* contacted over 100 manufacturers and research facilities in the United States, Europe, and Japan with visits to over 60 facilities to obtain first-hand information.

Their report contains the specification and application data for 69 cryocoolers in production at 28 manufacturers, 14 in the United States, eight in Europe, one in Israel, and four in Japan.

Figures 4.27 and 4.28, (reproduced from Smith *et al.* 1984) show the refrigeration capacity and specific power (power input, watts/watt refrigeration) as a function of the refrigeration temperature for the cryocoolers in commercial production. Figure 4.29 shows the weight of these systems as a function of the refrigeration capacity at operating temperatures of 4 K, 20 K, and 80 K

In addition to this most valuable repository of detailed technical information Smith *et al.* (1984) also attempted to survey cryocoolers

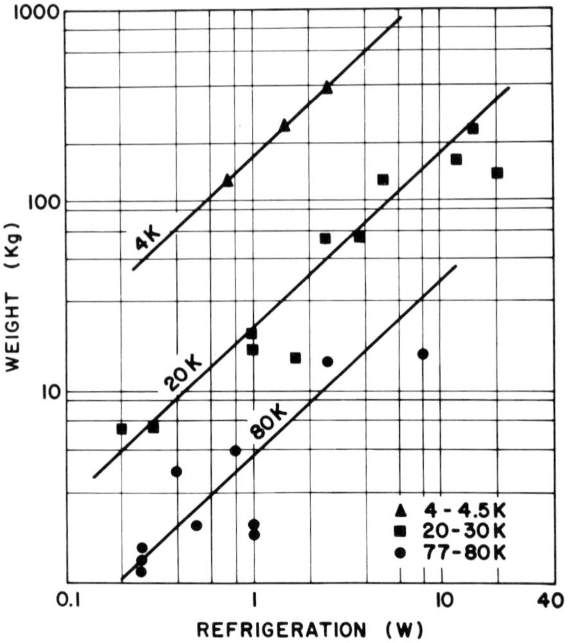

Fig. 4.29 Weight as a function of the refrigeration capacity for commercially available miniature cryocoolers (after Smith *et al.* 1984).

under development. Despite the obvious and pressing limitations of military classification and commercial confidentiality they were able to include some details of 32 miniature cryocoolers under development at 24 manufacturers or research institutes. Figure 4.30, reproduced from Smith *et al.* (1984), shows the refrigeration capacity and operating temperature of some of the cryocoolers in development.

The Smith report is valuable also as a directory of the names, addresses, and contact persons, most with telephone numbers, of 70 manufacturers of cryocooler systems or components, research institutes, sponsoring agencies, and some consumers of cryocoolers, 45 in the United States and Canada, 19 in Europe, one in Israel, and 14 in Japan. For the convenience of readers we have reproduced this directory in the appendix although the substantial interval since the report was prepared practically guarantees the personnel named will no longer be the principal contact.

Earlier, Strobridge (1974) of the National Bureau of Standards, Boulder, Colorado, presented another excellent survey of cryocoolers.

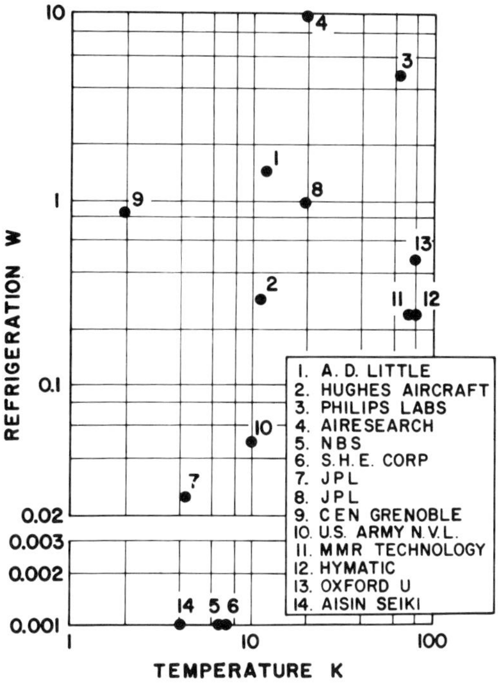

Fig. 4.30 Refrigerating capacity and operating temperature of some cryocoolers in development (after Smith *et al.* 1984).

The Strobridge survey was carried out on a much broader front than the later Smith survey which was limited to miniature cryocoolers. Strobridge presented charts, reproduced in Walker (1983), showing the efficiency (ratio of the actual coefficient of performance to the Carnot coefficient of performance), the volume, and the mass as a function of the refrigeration capacity for 144 different cryocoolers in three ranges: (a) 1.8–9 K, (b) 10–30 K, and (c) 30–90 K. He also provided another valuable chart showing the cost as a function of the input power for these 144 cryocoolers. Walker (1983) used the Strobridge charts to develop other charts showing the general mass and volume characteristics of cryocoolers as a function of the refrigeration temperature and capacity. The Strobridge 'experience' charts are most useful for planning purposes at the start of new projects involving cryogenic refrigeration using mechanical cryocoolers because they are based on such a broad range of actual achieved experience.

An extensive study of cryocoolers for spacecraft to provide long-term storage of cryogenic fluids was carried out by Jensen *et al.* (1971) for the NASA Manned Spacecraft Center, Houston, Texas. Another very comprehensive survey of cryogenic cooling techniques for space-borne systems was carried out by Donabedian (1972), well summarized and updated in Donabedian's excellent contribution to the *Infrared Handbook* (Donabedian 1978).

Various other surveys of cryogenic refrigerators for various specialist purposes were referenced by Walker (1983).

4.15 Closure

In this long chapter we have covered, albeit superficially, a good deal of ground in regard to mechanical cryocoolers, a field of extraordinary interest and diversity. Firstly we separated the various types of machines into two families:

(1) a family of mechanical refrigerators utilizing regenerative heat exchangers;

(2) a family of mechanical refrigerators not using regenerative heat exchangers.

Leaving the non-regenerative systems for consideration in a subsequent chapter we went on to consider the principal regenerative systems, classified by their cycle of operation as Stirling, Vuilleumier, Gifford–McMahon, Solvay, and pulse-tube refrigerators. For the purpose of providing an ideal yardstick to measure or compare performance we also considered the thermodynamically ideal refrigerator operating on the reversed Carnot cycle.

Following this survey of the elementary technology we proceeded to a comparison of the different systems noting the similarities and the differences. This was followed by a brief historical survey including contemporary developments and prospects for the future of mechanical cryocoolers of the regenerative genre.

References

Ackermann, R. A. and Gifford W. E. (1971). A Heat-Balance Analysis of a Gifford McMahon Cryorefrigerator. *Advances in Cryogenic Engineering*, **16**, 221. Plenum, New York.

Berry, R. (1993). Miniature Stirling Cryocoolers. Refrigerators for Cryogenic Sensors and Cold Electronics. Short Course Notes, University of California, Los Angeles.

Bush, V. (1938). Apparatus for Heat Transfer. US Patent No. 2,127,286. Aug. 16.

Chellis, F. (1978). Design Compromises in the Selection of Closed Cycle Cryocoolers. *Proceedings: Applications of Closed Cycle Cryocoolers to Small Superconducting Devices*. National Bureau of Standards Special Publications 508 (SupDoc. US Government Printing Office, Washington).

Chellis, F. and Hogan, W. (1964). A Liquid-Nitrogen Operated Refrigerator for Temperatures Below 77 K. *Advances in Cryogenic Engineering* (ed. K. Timmerhaus), **9**, 545–41, Plenum, New York.

Collins, S. C. and Cannaday, R. L. (1958). *Expansion Machines For Low Temperature Processes*. Oxford University Press.

Daunt, J. G. and Goree, W. S. (1969). *Miniature Cryogenic Coolers*. Office of Naval Research, NTIS AD 860866.

Davey, G. (1990). Review of the Oxford Cryocooler. *Advances in Cryogenic Engineering* **35B**, 1423, Plenum, New York.

Donabedian, M. (1972). *Survey of Cryogenic Cooling Techniques*. US Air Force SAMSO Report No. SAMSO-TR-73–74, AD No. 755 780, NTIS. Springfield, Va.

Donabedian, M. (1978). Cooling Systems. Chapter 15 in *The Infrared Handbook* (eds. W. Wolfe and G. Zissis), Infrared Information and Analysis Center, Environmental Research Institute of Michigan, Ann Arbor, Mich.

Gifford, W. E. (1966). The Gifford–McMahon Cycle. *Advances in Cryogenic Engineering*, **1**, 152, Plenum, New York.

Gifford, W. E. and Hoffman, T. E. (1961). A New Refrigeration System for 4.2 K. *Advances in Cryogenic Engineering* (ed. K. Timmerhaus), **6**, 82–94, Plenum, New York.

Gifford, W. E. and Longsworth, R. C. (1963). *Pulse-tube Refrigeration*. ASME Paper No. 63-WA-290.

Glaser, R. J., Ross, R. G. Jr., and Johnson, D. L. (1993). Cryocooler Tip Motion Suppression Using Active Control of Piezoelectric Actuators. *Proceedings of the Seventh International Cryocooler Conference* (ed. J. Ludwigsen), Kirtland AFB, Santa Fe, N.M.

Jensen, H. L., Nast, T. C., Glassford, A. P. M., Vernon, R. M., and Ekern, W. F. (1971). *Investigation of External Refrigeration Systems for Long-Term Cryogenic Storage*. Summary Report, Report No. NASA CR 115192, NASA Manned Spacecraft Center, NTIS Accession No. N71–37107 (Lockheed Missiles and Space Co., Sunnyvale, California) (February 1971), see also, same authors, same title: (a) System Review Report, NASA CR 115191, NTIS N71–37105 (May 1970); (b) Final Report, NASA CR 114920, NTIS N71–20279 (February 1971).

Jewell, C., Bradshaw, T., Orlowsha, A., and Jones, B. (1993). Present Life Testing Studies of Oxford Type Cryocoolers for Space Applications. *Proceedings of the Seventh International Cryocooler Conference* (ed. J. Ludwigsen), Kirtland AFB, Santa Fe, N.M.

Johnson, D. L., Mon, G. R., and Ross, R. G. Jr. (1993). Spacecraft Cooler Characterization. *Proceedings of the Seventh International Cryocooler Conference* (ed. J. Ludwigsen), Kirtland AFB, Santa Fe, N.M.

Kirk, A. (1874). On the Mechanical Production of Cold. *Proceedings of the Institute of Civil Engineering*, **37**, 244–315, London.

Lechner, R. A. and Ackerman, R. A. (1973). *Advances in Cryogenic Engineering*, **18**, 467.

Longsworth, R. C. (1967). *Advances in Cryogenic Engineering*, **12**.

Longsworth, R. C. (1971). A Modified Solvay-Cycle Cryogenic Refrigerator. *Advances in Cryogenic Engineering* (ed. K. Timmerhaus), **16**, 195–202, Plenum, New York.

Matsubara, Y., Gao, J., Tanida, K., Hireski, Y., and Kaneko, M. (1993). An Experimental and Analytical Investigation of 4 K Pulse Tube Refrigerator. *Proceedings of the Seventh International Cryocooler Conference* (ed. J. Ludwigsen), Kirtland AFB, Santa Fe, N.M.

Merkli, P. and Thomann, H. (1975). *Journal of Fluid Mechanics*, **70**, 161.

Mikulin, E. I., Tarasov, A. A., and Shkrebyonock, M. P. (1984). Low Temperature Expansion Pulse Tubes. *Advances in Cryogenic Engineering*, **29**, 620–9, Plenum, New York.

Minas, C. and Hualde, P. M. (1992). Dynamic Modelling of a Gifford–McMahon Cryorefrigerator. *Cryogenics*, **32** (7), 634–9, Butterworth-Heinemann, Oxford, UK.

Radebaugh, R. (1987). Pulse Tube Refrigeration: A New Type of Cryocooler. *Proceedings of the Eighteenth Conference on Low Temperature Physics*, Kyoto, Japan. *Journal of Applied Physics*, **26**, Supp. 26–3, 2076–81.

Ross, R. (1993). JPL Cryocooler Development and Test Program Overview. *Proceedings of the Seventh International Cryocooler Conference* (ed. J. Ludwigsen), Kirtland AFB, Santa Fe, N.M. Seven other papers on the JPL cryocooler work are included in the same proceedings.

Ross, R. and Johnson, D. L. (1993). Design and Test of a Comprehensive Facility for Life-Testing Space Cryocoolers. *Proceedings of the Seventh International Cryocooler Conference* (ed. J. Ludwigsen), Kirtland AFB, Santa Fe, N.M.

Schmidt, G. (1861). Theorie der Geschlossen Calorischen Maschine Von Laubroy und Schwatzkopff in Berlin. *Zeitshrift ver Oster Ing*, 79.

Senft, J. (1992). *Ringbom-Stirling Engines*. Oxford University Press, Newark, N.J.

Smith, J. L., Robinson, G. Y., and Iwasa, Y. (1984). *Survey of the State of the Art of Miniature Cryocoolers for Superconductive Devices*. Office of Naval Research Contract No. N0001483K0327, Cryogenic Engineering Lab., Department of Mechanical Engineering, Massachusetts Institute of Technology, Cambridge, Mass.

Sparr, S., Boyle, R., Cory, R., Conners, F., James, E., Fink, R., Arillo, V., and Manketoa, J. (1993). NASA/GSFC Cryocooler Test Program Results. *Proceedings of the Seventh International Cryocooler Conference* (ed. J. Ludwigson), Kirtland AFB, Albuquerque, N.M.

Strobridge, T. R. (1974). *Cryogenic Refrigerators—An Updated Survey*. National Bureau of Standards, Technical Note 655, Superintendant of Documents US Government Printing Office, 12 pp.

Swift, G. W. (1988). Thermoacoustic Engines. *Journal of the Acoustical Society of America*.

Taconis, K. W. (1951). Process of and Apparatus for Heat Pumping. U. S. Patent No. 2,567,454. Sept. 11.

Thirumaleshwar, M. and Subramanyam, S. V. (1986). Gifford–McMahon

Cycle: A Theoretical Analysis. *Cryogenics*, **26**, 177–88, Butterworth, Guildford, UK.

Thomas, W. H. (1873). British Patent No. 709. Producing Cold for Preserving Animal Food. (Science Museum Library, London).

Vuilleumier, R. (1918). Method and Apparatus for Inducing Heat Charges. US Patent No. 1,275,507. August 13.

Walker, G. (1980). *Stirling Engines*. Oxford University Press.

Walker, G. (1983). *Cryocoolers*. 2 vols, Plenum, New York.

Walker, G. (1989). *Miniature Refrigerators for Cryogenic Sensors and Cold Electronics*. Oxford Monographs on Cryogenics, Oxford University Press.

Walker, G. (1990). *Industrial Heat Exchangers: The User Guide*. 2nd edn. Hemisphere, Washington, D.C.

Walker, G. and Bingham, E. R. (1991). Micro and Nano Cryocoolers. *Proceedings of the Sixth International Cryocooler Conference* (ed. G. Green and M. Allen), David Taylor Naval Ship R. and D. Center, Annapolis, Md.

Walker, G. and Senft, J. (1984). *Free Piston Stirling Engines*. Springer Verlag, New York.

Walker, G., Weiss, M., Fauvel, R., and Reader, G. (1989). Microcomputer Simulation of Stirling Cryocoolers. *Cryogenics*, **29** (8) 846–9, Butterworth, Guildford, UK.

Walker, G., Reader, G., Fauvel, O. R., and Bingham, E. R. (1993). *The Stirling Alternative*. Gordon and Breach, Philadelphia, Pa.

Wheatley, J. and Cox, A. (1985). *Physics Today*, **38**, 8.

Wheatley, J., Hofler, T., Swift, G. W., and Migliori, A. (1985). An Intrinsically Irreversible Thermoacoustic Heat Engine. *Journal of the Acoustical Society of America*, **74**, 153–9.

White, R. (1976). *Vuilleumier Cycle Cryogenic Refrigeration*. AFFDL-TR-76–17, WPAFB, Dayton, Ohio, AD/A-027–055.

5

MECHANICAL CRYOCOOLERS.
PART 2: NON-REGENERATIVE SYSTEMS

5.1 Introduction

In this chapter we consider a family of mechanical cryocoolers different from those covered in the previous chapter. The machines of this chapter do not incorporate regenerative heat exchangers. In this family of non-regenerative cryocoolers there are three basic types. These are characterized by their thermodynamic cycle of operation and names of original investigators, specifically Linde–Hampson, Claude, and Joule–Brayton.

5.2 Principal features

The principal features of the three different systems are shown in Fig. 5.1. Diagram (a) at the left shows the Linde–Hampson system, diagram (b) the Claude system, and diagram (c) the Joule–Brayton system.

5.2.1 Commonalities

Figure 5.1 has been drawn to illustrate clearly the substantial commonality which exists among the three systems. For example, starting at the top of Fig. 5.1 and working down, all three systems include:

(1) a gas compressor;

(2) an aftercooler heat exchanger, usually water cooled;

(3) a recuperative heat exchanger;

(4) an expansion device;

(5) another heat exchanger to absorb the cooling load, the refrigerating effect.

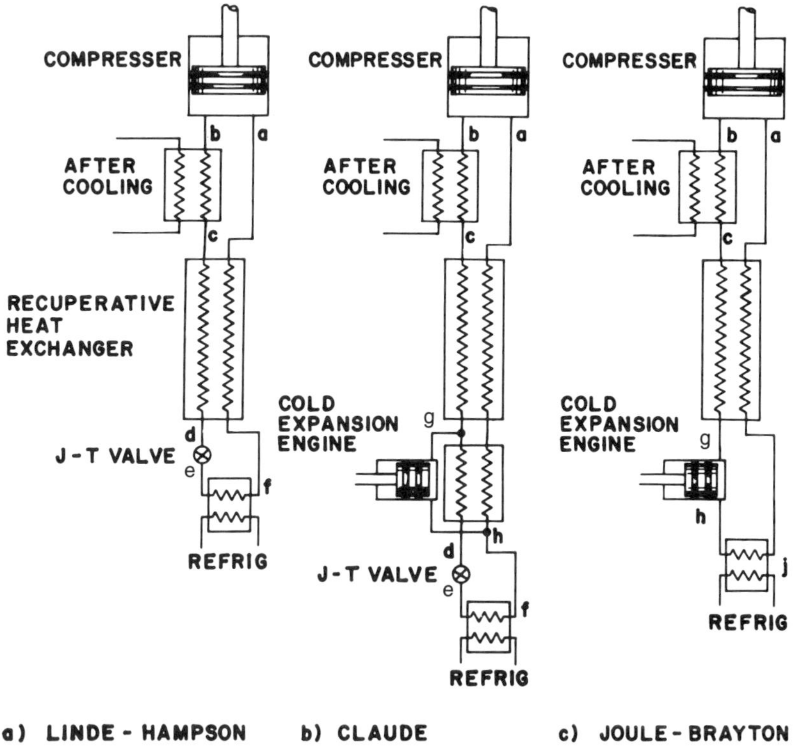

Fig. 5.1 Classification of non-regenerative mechanical cryocoolers.

5.2.2 *Differences*

The systems are different principally in the arrangements for expanding the gas to a low pressure and so producing the low temperatures necessary to achieve a cryogenic refrigeration effect.

These differences in the expansion arrangements are well illustrated by comparing the ideal thermodynamic cycles on the temperature–entropy planes. These are shown in Fig. 5.2 arranged as in Fig. 5.1; diagram (a) shows the Linde–Hampson cycle, (b) the Claude cycle, and (c) the Joule–Brayton cycle.

On all figures the process (a–b) shows an adiabatic compression of the gas from a low (ambient) pressure to a high pressure. Process (b–c) is the aftercooling of the compressed gas back to the ambient temperature.

Comparing now the Linde–Hampson and Claude systems we can see that both use a Joule–Thomson expansion nozzle to effect isenthalpic

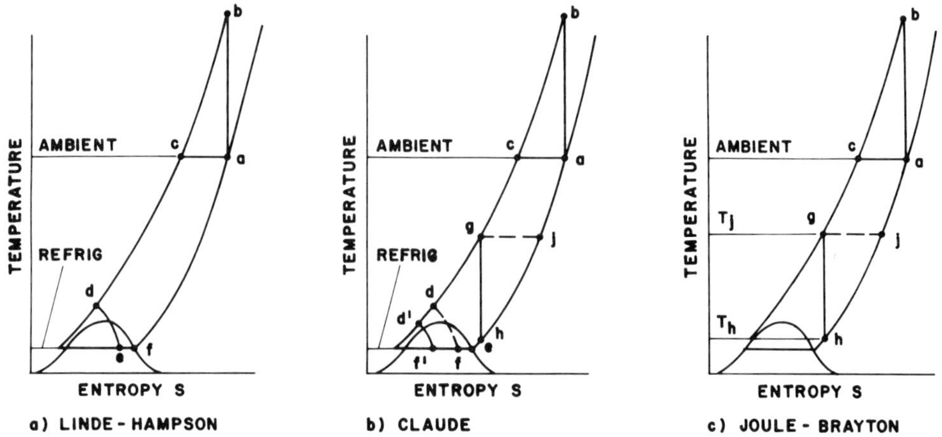

Fig. 5.2 Non-regenerative mechanical cryocoolers on the temperature–entropy plane.

expansion of the compressed gas, process (d–e) on Fig. 5.1(a) and 5.1(b). At the end of the expansion the fluid is partially liquefied (at state (e) but see the review of fundamentals in Chapter 3). The refrigeration effect, process (e–f), is the increase in enthalpy necessary to evaporate the fraction of working fluid that is liquefied to the dry saturated condition at (f).

Now let us compare the Claude and Joule–Brayton cycle systems. Both incorporate an expansion engine to expand the gas isentropically to the lower pressure, process (g–h) on Fig. 5.1(b) and (c).

It is clear the Claude cycle is singular in that it *incorporates both isentropic and isenthalpic expansion of the gas* to the lower pressure. To accomplish this dual expansion process the flow of compressed fluid passing in the recuperative heat exchanger is split at stage (g), Fig. 5.2(b). Part of the flow proceeds thence to the expansion engine, the remainder continuing in the recuperative heat exchanger for further cooling to state (d) where it expands isenthalpically to state (e) and evaporates to state (f) in performing refrigeration, process (e–f). After partial reheating to state (h) the low-pressure fluid from the Joule–Thomson expansion mixes with the cold low-pressure fluid exhausting from the expansion engine at the end of isentropic expansion. The combined flows then return through the recuperative heat exchangers to the initial state (a) at inlet to the compressor.

It is interesting to note that in the Claude cycle the fraction of fluid flowing in the expansion engine does not contribute directly to the refrigerating effect. Instead the refrigeration produced by the expansion engine is consumed to enhance the cooling achieved in the lower-stage recuperative heat exchanger.

In this way the fraction of compressed gas cooled in the lower-stage recuperator (process (g–d)), Fig. 5.2(b), *en route* to the Joule–Thomson expansion achieves a lower temperature state (d′) than it would if there were no expansion engine. This reduction in the temperature immediately prior to Joule–Thomson comparison of the process (d′–f′) on Fig. 5.2(b) with the process (d–f) starting at a higher temperature.

Of course it must be remembered that part of the total fluid compressed has been diverted to the expansion engine. Therefore, although the *fraction* of fluid expanding in the Joule–Thomson valve that is liquefied is increased the actual mass rate of flow passing to the Joule–Thomson nozzle is reduced. Clearly, therefore, there are obvious optimization studies that must be carried out to determine the best division of fluid between the isenthalpic and isentropic routes. Further details of these 'trade-off' studies were given by Barron (1984) and by Walker (1983).

The useful refrigerating effect of the Joule–Brayton cycle is achieved during the constant-pressure heating process (h–j), Fig. 5.2(c). Cold gas at the end of the isentropic expansion at state (h) is heated along the constant-pressure line to state (j) increasing in temperature from a minimum of T_h to the temperature T_j at which it enters the recuperative heat exchanger *en route* to the compressor inlet.

In many applications refrigeration at a constant temperature is required. In such cases the change in temperature from T_h to T_j in the Joule–Brayton cycle is thermodynamically unattractive. The minimum temperature at which cooling can be supplied, at constant temperature, is the maximum temperature of the cooling process, temperature T_j. It is clearly thermodynamically inefficient to have cooling available at a much lower temperature but be unable to use it at that lower temperature. In other applications it is often advantageous to have cooling available over the range of temperatures T_h to T_j. Many Joule–Brayton systems utilize high-speed rotating expanders and compressors with continuous flow so that continuous steady-state cooling effects are obtained.

The refrigeration effect of both the Linde–Hampson and the Claude cycle is obtained at constant temperature as the partially liquefied gas evaporates at constant pressure, process (e–f). It will be recalled from Chapter 3 that on the temperature–entropy plane evaporation and condensation processes occur at constant temperature and pressure within the liquid vapour envelope.

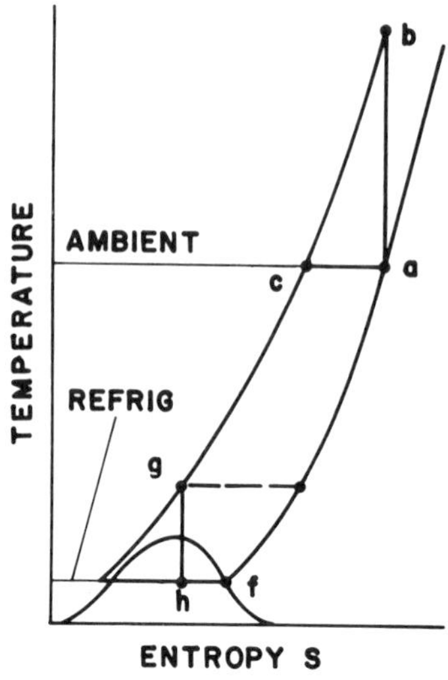

Fig. 5.3 Joule–Brayton cycle with a 'wet' expansion.

5.2.3 Wet-expansion engine

Figure 5.2 is drawn so the fluid at the end of the isentropic expansion, state (h) in both the Claude and Joule–Brayton cycles, is outside the liquid vapour envelope. In other words none of the gas has been liquefied in the expansion. This is not a necessary thermodynamic condition. There is no fundamental reason why the gas cannot be at least partially liquefied during the expansion, as shown by the Joule–Brayton cycle in Fig. 5.3. Here the expansion process (g–h) does penetrate the liquid–vapour envelope, some of the gas is liquefied and the refrigerating effect, process (h–f) does occur at constant temperature T_f. This goes a long way to overcoming the problem outlined above that, in the normal Joule–Brayton cycle, the refrigeration is available at a varying temperature.

An expansion of the type shown in Fig. 5.3, process (g–h), is called a *wet expansion*. It will be recalled from Chapter 3 that a fluid at state (h) is a theoretical concept not feasible in practice. Fluid at this conceptual state (h) is actually a mixture of two distinct fluids, one saturated liquid at state (k), the other saturated liquid at state (f) with physical properties that are quite different. The saturated liquid, with a high density, may be dispersed as small droplets dispersed in a homogeneous stream of saturated vapour.

Such a mixed stream of fluids is difficult to handle with rotating equipment, particularly the very high-speed turbo-expanders used for low-capacity Joule–Brayton systems. The complex blade angles and curvatures necessary to minimize friction effects with the saturated vapour are not suitable for the small, bullet-like, droplets of the saturated liquid and so the erosion and wear of the turbine blades occurs sometimes catastrophically. Moreover frictional effects in fluid flow are, in general, a function of the square of the velocity of the fluid. Doubling the fluid velocity increases the friction loss four times!

Wet expansion processes can be more readily accomplished with reciprocating expansion engines. These operate much slower and consequently are much larger than turbo systems. Moreover, the geometry of reciprocating expanders is nothing like so critical as turbo machines because the velocities involved are orders of magnitude less.

5.3 Theoretical analysis of non-regenerative mechanical cryocoolers

The elementary analysis of recuperative systems is relatively simple compared with regenerative systems. This is because the recuperative systems are usually continuous-flow, steady-state units whereas regenerative systems are always unsteady-state, continuously variable flows of different velocities, pressures, densities, and temperatures, with continuously variable conditions.

Recuperative system analysis is best approached by use of the general energy equation for steady flow. Consider the thermodynamic 'black box' shown in Fig. 5.4 with a variety of energy inputs to the box, including kinetic energy, internal energy, flow work, and heat transfer. There are also a variety of energy outputs from the box. When the system is in steady-state operation, the energy input must be equal to the energy output. If this equality were not so, an accumulation or depletion of energy would occur and the system would not be in steady state.

Summarizing the energy inputs and equating these to the energy outputs we can express the general steady-flow energy equation in mathematical terms as:

$$\dot{m}Z_1 + \dot{m}C_1^2 + \dot{m}U_1 + p_1V_1 + Q = \dot{m}Z_2 + mC_2^2 + \dot{m}U_2 + p_2V_2 + W \quad (5.1)$$

where $\dot{m}$ is the mass rate of flow of the fluid, Z is the height of fluid above some prescribed datum, C is the velocity of the fluid, U is the internal energy of the fluid, p is the pressure of the fluid, V is the

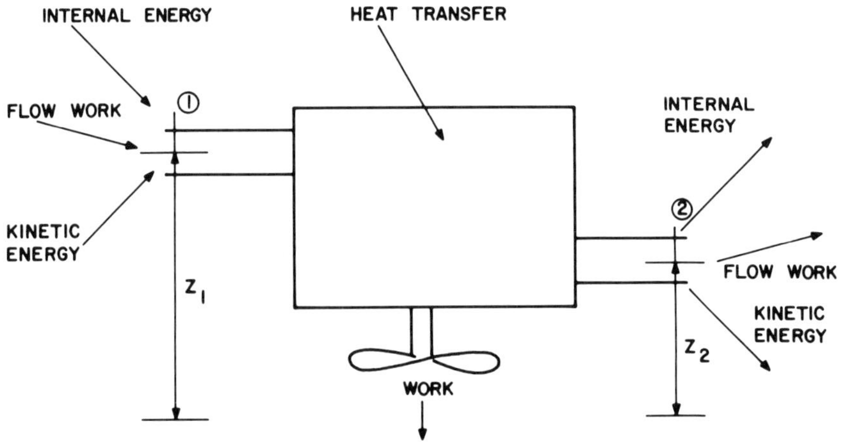

Fig. 5.4 Steady-flow thermodynamic 'black box'.

volume of fluid flow, Q is the net heat transfer to the fluid, and, W is the net work output. Frequently the terms $(\dot{m}U + pV)$ can be combined to the new term mH where H is the enthalpy of the fluid.

In many cases the change in the datum heat $[\dot{m}(Z_1 - Z_2)]$ is negligibly small. Then Eq. (5.1) can be restated as:

$$\dot{m}(H_1 - H_2) = \dot{m}(C_2^2 - C_1^2) + W - Q . \quad (5.2)$$

The change in enthalpy from state 1 to state 2 is therefore the sum of the change in kinetic energy, the work done, and the heat transferred.

Elementary analysis of recuperative cryocoolers can be approached by considering enthalpy values at the different locations of the system. Consider the Joule–Brayton cycle system shown in Fig. 5.5. The change in enthalpy per unit mass flow $(H_b - H_a)$ during compression is the work of compression that must be supplied to the cycle. This assumes the velocity of the gas at state (b) is about the same as at state (a). The heat transfer Q during compression is zero because the process is assumed to be isentropic (i.e. $\Delta S = \Delta Q/T = 0$).

Similarly, the heat transfer $Q_{(b-c)}$ from the high-pressure fluid in the water-cooled heat exchanger is $(H_b - H_c)$ and in the recuperative heat exchanger $Q_{(c-g)}(H_c\text{-}H_g)$.

The work produced in the expander $W_{g-h)}$ is $(H_g - H_h)$ and may be applied to partially offset the work of compression but in many cases is simply dissipated by a friction brake or electrical generator feeding a resistive grid.

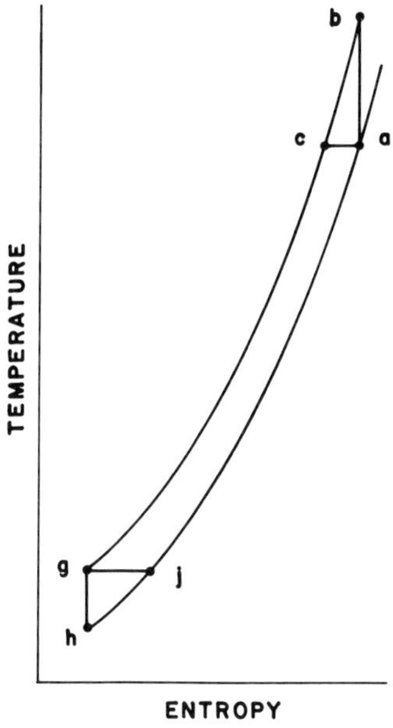

Fig. 5.5 Joule–Brayton cryocooler.

Finally the useful refrigeration of the cycle is $Q_{(h-j)} = (H_h - H_j)$ and the heat transfer in the recuperative exchanger is $Q_{(j-a)} = (H_j - H_a)$. Ideally the heat transferred to the low-pressure fluid in process (j–a) is equal to the heat transferred from the high-pressure fluid in process (c–g), i.e. $Q_{(c-g)} = Q_{(j-a)}$. In practice this equality does not hold and the heat gained by the low-pressure fluid is less than the heat lost by the high-pressure fluid. The ratio $Q_{(j-a)}/Q_{(c-g)}$ is termed the *effectiveness* of the recuperative heat exchanger.

The net work input to the cycle, W_n, (assuming the expander work output is used to reduce the compression work) is:

$$W_n = (H_b - H_a) - (H_g - H_h) . \tag{5.3}$$

The heat extracted in refrigeration is:

$$Q_C = (H_h - H_j) . \tag{5.4}$$

Therefore the coefficient of performance is:

$$\text{COP} = \frac{Q_C}{W_n} = \frac{(H_h - H_j)}{(H_b - H_a) - (H_g - H_h)} . \tag{5.5}$$

Of course there are many effects that need also to be considered including aerodynamic friction effects in the compressor and expander (manifest in the isentropic efficiencies for both compressor and expander) and frictional pressure drops which reduce the expansion pressure ratios and thus the amount of work recovered. Other important practical effects involve inefficiencies of the heat exchanger process arising from limited heat transfer, perhaps inadequate heat transfer area, thermal conduction from the ambient temperature parts to the cold parts, aerodynamic frictional pressure drop, and reheating effects in the heat exchanger, and so on.

5.4 Constituent elements of non-regenerative mechanical cryocoolers

It is clear from the foregoing that the essential constituent elements of non-regenerative mechanical cryocoolers are:

(1) compressors;

(2) expanders;

(3) heat exchangers.

We shall consider these component parts separately then review some ensembles of the elements arranged in particular forms. Some of these have gone into commercial service. Others continue as research development topics with potential for application in the near term.

5.4.1 Compressors

Nomenclature Compressors can be divided broadly into two groups:

(1) positive displacement compressors;

(2) turbo-compressors.

Many positive-displacement compressors have reciprocating pistons in cylinders. Other rotary types have sliding vanes in an eccentric rotor or in other cases mating lobes arranged in helical fashion on twin contra-rotating rotors. They are provided with inlet and outlet valves or other features to regulate the admission of low-pressure fluid and delivery of high-pressure fluid to and from the compressor. They are characterized by relatively low flow rates, relatively slow speeds of operation, and the ability to function with very high pressure ratios, i.e. p_{max}/p_{min} of 200 or even 400 to 1 in machines with up to four stages of compression.

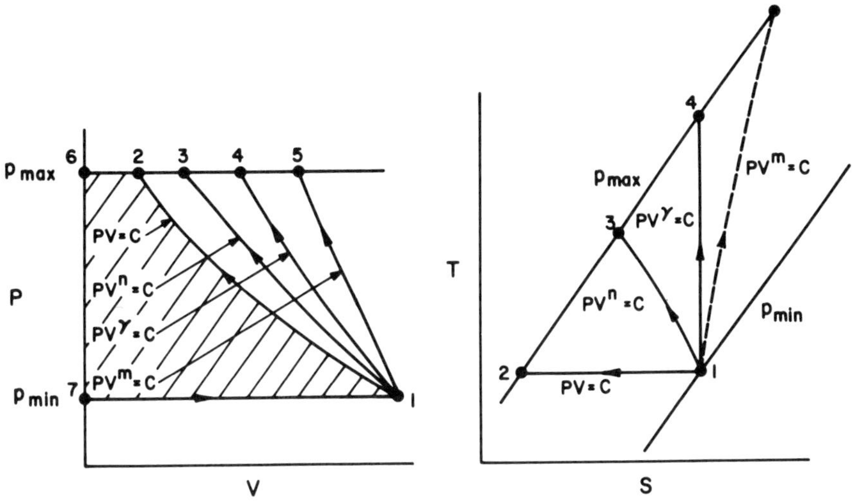

Fig. 5.6 Compression processes on the $P-V$ and $T-S$ planes.

Turbo-compressors are not positive-displacement machines and depend on aerodynamic effects to coax fluid from a low pressure to a high pressure. This is usually accomplished by first accelerating the fluid then converting the velocity energy to pressure energy by means of a diffuser. Turbo-compressors characteristically operate at very high speed with high flow rates and low pressure ratios, $p_{\max}/p_{\min}$ rarely exceeding 5 or 6 to 1 even with many (12 or more) stages of compression.

Figure 5.6 illustrates, on the $P-V$ and $T-S$ planes, the different processes for compression.

1. Process (1–2) is an isothermal compression represented by PV = const.
2. Process (1–4) is an isentropic compression represented by $pV^{\gamma} =$ const., where γ is the ratio of specific heats of the gas, $\gamma = C_p/C_v$.
3. Process (1–3) is a polytropic process between an isothermal and an isentropic process. The compression can be represented by $pV^n =$ const., where $(1 \leqslant n \leqslant \gamma)$.
4. Process (1–5) is polytropic represented by the equation $pV^m =$ const., where $m \geqslant \gamma$.

The ideal reference cycle for reciprocating compressors is the cycle with isothermal compression. In a reciprocating compressor the gas passes through the machine at a relatively slow pace and so some cooling is possible. To actually achieve isothermal compression would require either:

(1) infinite rates of heat transfer between the gas in the cylinder and the cylinder walls; or

(2) a speed of operation so low as to be impractical.

In practice reciprocating compressors follow the intermediate polytropic path, process (1–3) on Fig. 5.6.

For rotary turbo systems the ideal reference cycle is the isentropic compression process. In a rotary machine at high speed the gas particles are moving so fast that there is simply no time for cooling to take place. Moreover, there is some frictional heating as the high-speed flow passes over the blades so that in practice rotary compressors follow the polytropic process (1–5).

The *isothermal efficiency* for a compressor (usually limited to reciprocating compressors) is defined as

$$\eta_{(\text{iso})} = \frac{\text{isothermal compresssion work}}{\text{actual compression work}} = \frac{\text{area 1–2–6–7}}{\text{area 1–3–6–7}} = \frac{H_2 - H_2}{H_3 - H_2}\,. \tag{5.6}$$

The *isentropic efficiency* for a compressor (usually rotary but sometimes also used for reciprocating) is defined as

$$\eta_{(\text{isen})} = \frac{\text{adiabatic compresssion work}}{\text{actual compression work}} = \frac{\text{area 1–4–6–7}}{\text{area 1–5–6–7}} = \frac{H_4 - H_1}{H_5 - H_1}\,. \tag{5.7}$$

Since both the isentropic and isothermal efficiencies are arbitrary ratios it is not possible to say one is 'better' in any way than the other. The isentropic work is the ideal minimum work conceivable in a high-speed turbo system whereas the lesser, isothermal work is conceivably possible in the slower reciprocating machines.

Enthalpies, H, can be calculated or simply read from the appropriate point on the $T–S$ or $H–S$ diagram. Alternatively, the areas can be measured on the $P–V$ work diagram. The *compression work* measured by areas on the $P–V$ plane or changes in the enthalpy, is called the *indicated work*. It is considerably less than the *actual* or *brake work* required to drive the compressor. Additional work over and above the indicated work is necessary to overcome mechanical and aerodynamic friction losses, pneumatic losses in charging (process (7–1) in Fig. 5.6) and discharging (process (6–2)), and for driving accessories.

The ratio of indicated work to actual shaft input is called the *mechanical efficiency* of the compressor:

$$\eta_{(\text{mech})} = \frac{\text{indicated work}}{\text{brake work}} . \quad (5.8)$$

The mechanical efficiency usually ranges from 60–80 per cent for reciprocating engines and up to 90 per cent for turbo systems. This is quite different and should not be confused with the forementioned isothermal or isentropic efficiencies whose values may range from 40 to 85 per cent. The term 'efficiency' is often loosely applied in the literature, particularly manufacturers literature, so it is frequently necessary to enquire and confirm precisely how the terms are defined and what is being described.

Reciprocating compressors have erankshafts with main bearings, connecting rods with big and small end bearings, pistons with piston guide rings, and sealing rings. They may also be equipped with water cooling requiring some pump work, with forced lubrication requiring oil pump work, and perhaps with a fan for air cooling. Therefore, the mechanical and hydraulic losses tend to be appreciable compared with the pneumatic effects arising from the relatively slow-moving gas being compressed.

On the other hand, rotary systems have only main shaft bearings and rotary shaft seals which consume little power. The principal losses are, therefore, pneumatic arising from friction and windage effects of the high-speed rotation of the disks, drums, and blades within the compression space.

Frequently the isentropic efficiency is quoted for reciprocating compressors rather than isothermal efficiency. This is adopted by compressor manufacturers to put their machines in a better light since the ratio of the isentropic work to the actual shaft input is larger than the ratio of isothermal work to actual shaft input. Both are arbitrary ratios and the actual work required remains the same regardless of the efficiency used. However, because of this situation it is important when assessing compressor efficiencies to establish if the efficiency being quoted is isentropic, isothermal, or mechanical.

Yet another efficiency that is much used, the *volumetric efficiency*, is defined as:

$$\eta_{(\text{vol})} = \frac{\text{volume of gas actually pumped}}{\text{displacement volume of compressor}} . \quad (5.9)$$

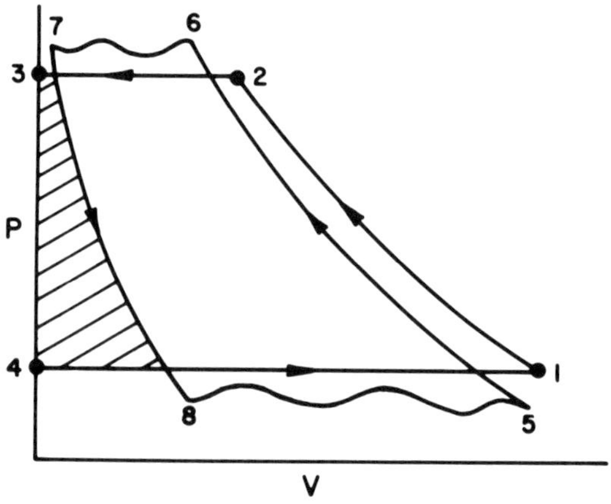

Fig. 5.7 Comparison of actual and ideal compressor work diagram.

The volume of gas passing through the compressor is always less than the piston displacement volume. The reduction is due to a number of causes illustrated by the work diagram shown in Fig. 5.7. This shows the ideal work diagram 1–2–3–4 superimposed on the actual compressor indicator diagram. The effects have been exaggerated for the sake of clarity. Due to pressure losses in the intake duct (perhaps including a filter) the pressure at the start of compression (5) is less than the theoretical value (1).

The mass of gas in the cylinder can be calculated as:

$$m_c = \frac{p_5 V_5}{R T_5} \tag{5.10}$$

where m_c is the mass of gas in the cylinder, p_5, V_5 and T_5 are the pressure, volume, and temperature at point (5), and R is the characteristic gas constant. Clearly a reduction in the pressure will mean a reduction in the mass of gas contained.

Now the actual compression process follows the path (5–6) rather than the ideal path (1–2). Moreover, it is necessary to compress the gas to a somewhat higher final pressure, (6), than theoretically required, (2), so as to be able to overcome pressure losses in the delivery valve and ductwork, including the aftercooler. This has no effect on the volumetric efficiency but requires extra work and so reduces isothermal or isentropic efficiency.

The principal effect on volumetric efficiency arises from residual gas remaining in the cylinder at the end of the delivery of compressed gas (7). It is not possible for the compressed fluid to be discharged completely from the compressor cylinder. A small amount of residual fluid (2–5 per cent) will always remain in valve cavities and clearance spaces when the piston reaches top dead centre position. Now as the piston descends the compressed residual gas expands from (7) to (8) and no fresh charge will be inhaled to the cylinder until the pressure falls to (8) and the inlet valve opens to admit new gas.

A sizeable fraction of the piston stroke is thus consumed in this re-expansion of the compressed gas and is the biggest factor reducing the volumetric efficiency. Expansion of the clearance gas does work on the piston thus returning the compression work expanded in compressing it. Therefore the net work input to the compressor is decreased proportionately as the net mass flow entering the compressor is decreased because the volumetric efficiency is less than 100 per cent as the result of clearance effects.

A reduction in inlet pressure or increase in inlet temperature will also reduce the mass flow to the compressor. When comparing one compressor with another on the basis of volumetric efficiency or the 'free air delivery rating' it is important to specify the same reference pressure and temperature. For comparative purposes it is often best to reduce the available data to terms of *mass rate of flow* at the *standard temperature and pressure* (STP) of 0 °C and 101.3 kPa.

Multistage compression When gases are compressed over a large pressure range it is advantageous to perform the compression in two or more stages with intercooling between stages and aftercooling the final stage. This is illustrated in Fig. 5.8 where the $P-V$ and $T-S$ diagrams for two-stage compression are superimposed on those for a single stage.

The single-stage compression diagram is represented by process (1–11–6–10). The two-stage compression process is represented by the low-pressure stage, process (1–2–8–9), and the high-pressure stage, process (3–4–6–7). At the intermediate pressure the fluid is assumed to be cooled, process (2–3), to the original temperature. Process (1–3–5) is the isotherm for the temperature at state 1.

The advantage of multistage compression is twofold. It increases the volumetric efficiency and also the isothermal (or isentropic) efficiency. Increase in the volumetric efficiency comes about because in a two-stage machine the clearance gas expands only to (9) rather than the larger volume, (10), with single-stage compression. The low-pressure stage can therefore admit a greater volume of new gas.

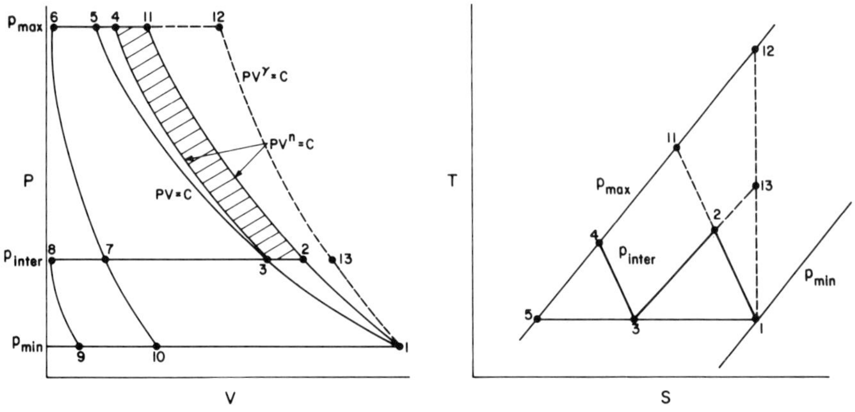

Fig. 5.8 Single and multistage gas compression.

The isothermal or isentropic efficiency increases because of the reduced compression work required in the second stage when inter-cooling is used between the stages to reduce the temperature back to the initial temperature. The saving in work is indicated by the shaded area (2–11–4–3) in Fig. 5.8.

The disadvantage of multistage compression is that a more complicated and expensive machine is required with more moving parts, bearings, pistons, seals, valves, and intermediate pressure heat exchangers and intercoolers. Both the savings in compression work and the complications and costs increase with the number of stages. However, the benefits of additional stages of compression occur at a progressively diminishing rate so it is rarely necessary or worthwhile to exceed four or perhaps five stages even for the highest pressure ratios.

To a first approximation the optimum arrangement for a multistage compressor is to have equal compression work or equal pressure ratios in each stage, that is:

$$P_r = n\sqrt{\frac{p_{\max}}{p_{\min}}} \tag{5.11}$$

where p_r is the pressure ratio per stage, n is the number of stages, $p_{\max}$ is the maximum pressure (at the end of compression), and $P_{\min}$ is the minimum pressure (at the start of compression). Work diagrams for a practical two-stage compressor are drawn in Fig. 5.9.

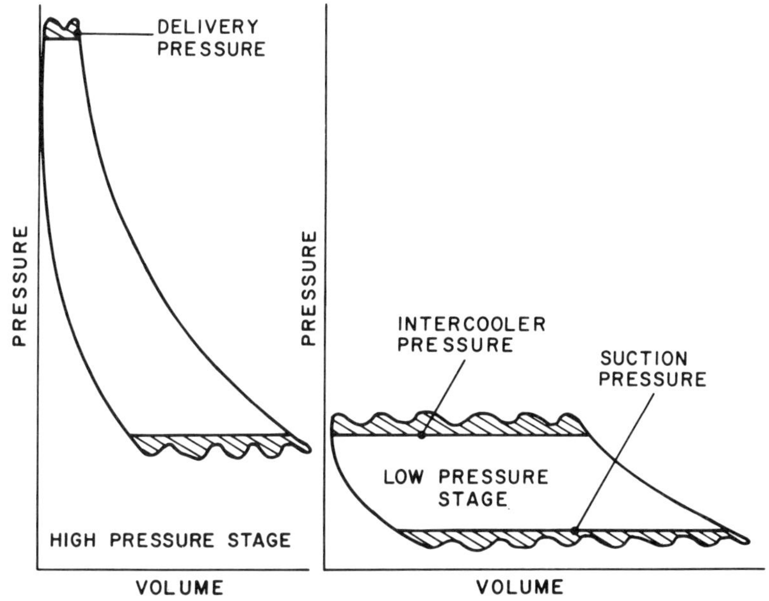

Fig. 5.9 Work diagrams for a practical two-stage compressor.

Compressors in Linde–Hampson cryocoolers Effective and reliable gas compressors are a critical element in Linde–Hampson cryocoolers. For the miniature systems used in infrared night vision and missile guidance systems there are three approaches:

1. Use relatively large, high-capacity air compressors, with sophisticated gas cleaning equipment to charge storage bottles to very high pressure. This is the most satisfactory method of producing compressed air of satisfactory purity. However, the logistics problems involved in delivering bottles to the use location and retrieving depleted bottles is not attractive.

2. Use smaller, low-capacity, open-cycle air compressors mounted on the vehicle or aircraft at the 'point of use', customarily to charge air storage bottles rather than direct use in the cryocooler. Generally the minicooler and its associated thermal sight is to be used at night during 'silent watch' so that quiet operation from bottle storage is preferred to the noise of an electrically powered compressor. The bottles depleted overnight can be recharged the next day when the noise of the compressor will be drowned by the high levels of daylight background noise.

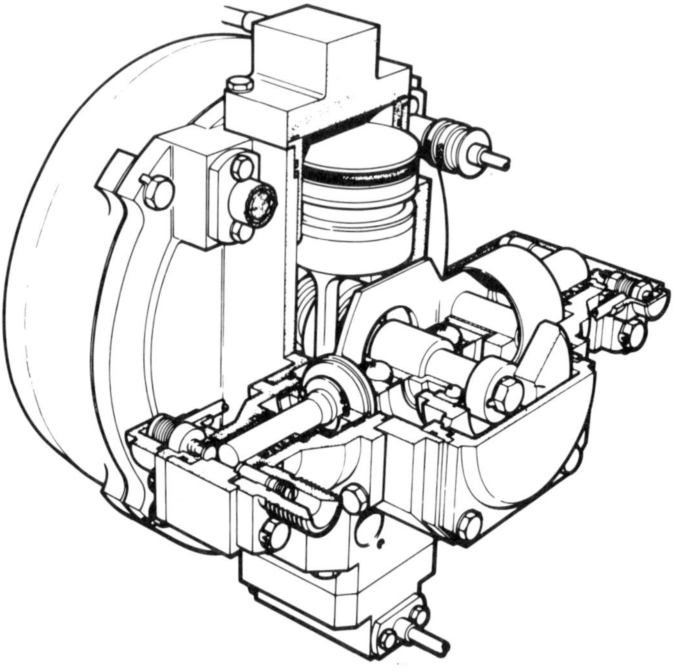

Fig. 5.10 Four-stage miniature gas compressor (courtesy Hymatic Engineering Company Ltd).

3. A miniaturized, closed-cycle gas compressor can be used, operating continuously and supplying one or more Joule–Thomson nozzles with compressed gas. Such a unit can operate with air in an open-cycle mode or, in special situations, on a closed-cycle with other gases, such as argon, Freon, or neon. Figure 5.10 shows a miniature reciprocating gas compressor of this type manufactured by the Hymatic Engineering Company Ltd. of Worcester, England. It was expressly designed for use on military vehicles. It is a four-stage normally aspirated machine with dry lubricated seals and grease-packed bearings to allow use in any attitude and which delivers compressed gas, usually air, at 200 atm. The delivery flow is adjusted to match the minicooler requirement by variation in the compressor speed. The compressor space envelope is 150 mm × 160 mm × 175 mm and it weighs 5.5 kg. The gas cleaner is 20 mm diameter × 140 mm long and has a cleaning life in excess of 400 hours. At start-up when the minicooler has maximum demand for gas the compressor runs in excess of 2500 revolutions per minute. When the cool-down is

complete and the cooler is operating at the maintenance condition the compressor speed is about 200 revolutions per minute and the noise level is 62 dB(A) at 1 m.

Cooling The operation of any compressor requires an input of mechanical work. This is eventually converted to heat, principally the heat of compression of the gas and also the mechanical and aerodynamic friction resulting from operating the compressor. In turbo-compressors the pressure ratio is relatively low so the rate of work input per unit of gas mass flow is much lower and so the heating effects are much less than for reciprocating compressors. In the slow-speed, low mass flow reciprocating systems the heat of compression must be removed to prevent the build-up of high temperatures with consequent mechanical damage following breakdown of the lubrication system or dry rubbing elements.

Both liquid-cooled and air-cooled compressors are found in practice. Water cooling is very much more effective than air cooling. However for the small very compact systems used on vehicles and aircraft, air cooling is preferred. The water or other liquid coolant used for a water-cooled compressor must itself be cooled, usually with an air-cooled convective heat exchanger similar to an automotive radiator. This cooling device is often as large as the compressor itself. The addition of liquid cooling does tend to reduce the temperature of operation. This has a beneficial effect on the volumetric efficiency of the compressor and also decreases, often dramatically, the noise level of the compressor. Cooling of the compressor drive motor should not be overlooked. Large electric motors have a very high conversion efficiency (sometimes exceeding 90 per cent) but for small motors the conversion efficiency is much less, 50 per cent or so in miniature machines.

Gas clean-up Clean-up of the compressed gas produced from reciprocating or other types of positive-displacement compressors is usually required. It is rarely necessary with turbo-compressors and never to the same degree as with reciprocating systems because the pressure ratios involved are so much less.

Gas cleaners are used to remove water vapour, carbon dioxide vapour, and other contaminants, such as oil and wear debris after compression.

Oil lubrication of the compressor running gear, cranks, connecting rods, and so on, in the crankcase is very attractive to minimize mechanical friction and wear. However, it is almost impossible in compact machines to prevent contamination of the compressed gas output by leakage of oil into the compressor cylinder. Special measures to prevent this contamination are necessary in oxygen compressors (to eliminate the possibility of spontaneous combustion or explosion) but are not practical for most other types of compressor. The contaminants must subsequently be recovered from the compressed gas if clean contaminant-free gas is required, as in miniature Joule–Thompson expansion.

In some types of compressor, for example helical screw compressors, a type of relatively high-speed, positive-displacement compressor, oil is deliberately added to the suction air or injected to the compressors to improve sealing and cooling of the air as well as to lubricate the moving parts.

Gas clean-up equipment generally comprises a water and oil separator, perhaps using centrifugal action or some sudden reversal of direction to separate the larger liquid droplets from the gas flow. This would likely be followed by one or more stages of filter or molecular sieve, usually in replaceable cartridge form. The size of the cartridge dictates the maintenance interval at which the cartridges must be replaced.

Some systems incorporate regenerative gas cleaners controlled by a small microprocessor. These regenerative units include two separate systems each containing a water/oil trap and a molecular sieve filter. One column is used to clean up the compressed air or gas while the other is being regenerated. Regeneration is accomplished by heating and reverse flow of gas through the system to exhaust so as to remove the contaminants. In start-up/cool-down operation when the gas flow is highest it may be possible to operate with the regenerative systems in parallel, thereby duplicating the capacity, until the demand eases and the motor/compressor speed reduces to the 'cruise' condition.

5.4.2 Expanders

Isentropic versus isenthalpic expansion All mechanical cryocoolers involve expansion of the working fluid from a high pressure to a low pressure. The working fluid is usually a gas, but can be a liquid (gas previously liquefied) or wet vapour (gas cooled below the critical temperature with liquid droplets entrained in it).

The expansion process can approximate various idealized thermodynamic processes, namely, isothermal, isenthalpic, isentropic, and polytropic. These are represented on the temperature–entropy diagram (T–S plane) as shown in Fig. 5.11.

For all four processes we will assume that initially the gas is at some high pressure and temperature T_1. The gas expands to a low pressure and then heat is supplied to bring the temperature back to the initial temperature T_1. An apparatus to carry out such comparative work is shown in Fig. 5.12. It consists of a piston reciprocating in a cylinder closed at one end. A valve is connected to the cylinder through which

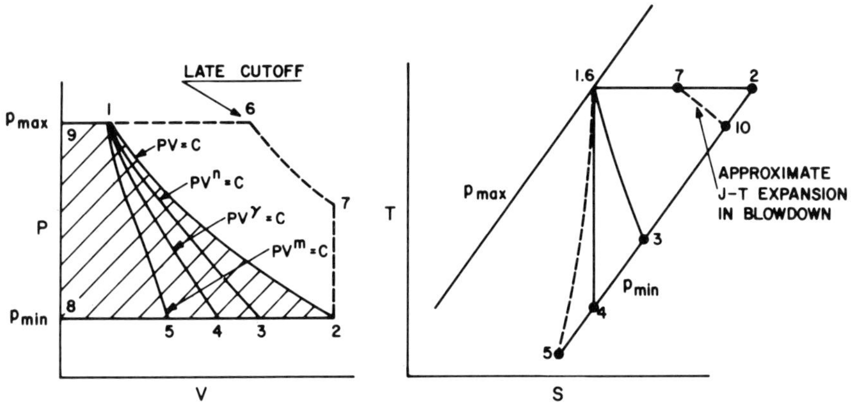

Fig. 5.11 Expansion processes on the temperature–entropy plane.

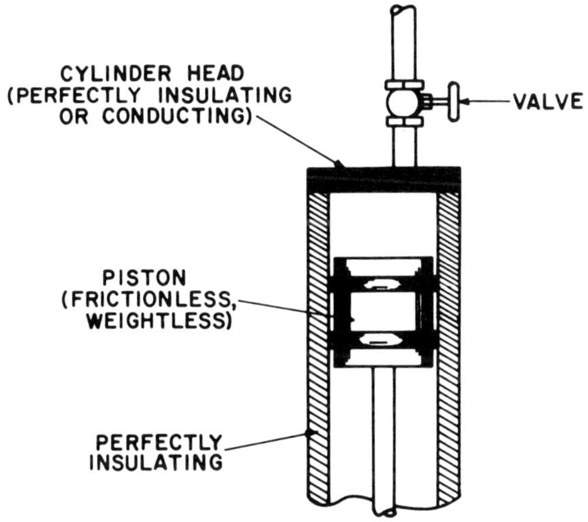

Fig. 5.12 Idealized expansion engine.

fluid is admitted or expelled from the cylinder. The cylinder head is assumed to be perfectly insulating or perfectly heat conducting as required. The walls of the cylinder are perfectly insulated and the piston is assumed to be without mass and to experience no friction in moving about the cylinder.

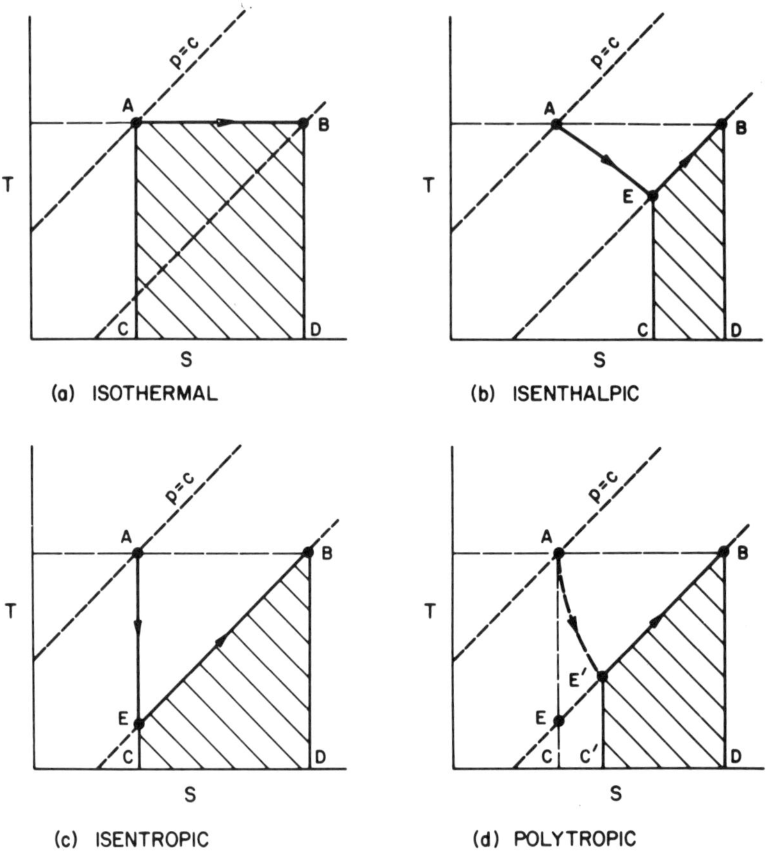

Fig. 5.13 Temperature–entropy diagrams for expansion processes.

For the isothermal process represented in Fig. 5.13a assume that the cylinder is filled with gas at condition A. The filling process occurs when the piston is at the top of its stroke and the volume between the piston and cylinder head is least. The inlet or fill valve is closed when the piston moves down the cylinder and the gas expands to a low pressure. For the isothermal expansion the cylinder head is assumed to be perfectly conducting and, as the gas expands, the temperature tends to decrease so that heat is transferred through the cylinder surroundings to maintain the temperature constant. The heat abstracted from the cylinder surrounds, the refrigeration effect, is represented by the shaded are A–B–C–D in Fig. 5.13a. Work is done on the piston by the gas.

The amount of work is exactly equal to the heat abstracted from the surroundings.

For the isenthalpic process represented in Fig. 5.13b by process (A–E) assume that, with the cylinder empty of gas, the piston is at the bottom of the stroke with the volume between the piston and the cylinder head at its maximum value. The inlet valve is cracked open slightly and working fluid is slowly admitted to the cylinder. The pressure decreases and eventually reaches the final required pressure. Gas passing the inlet valve is said to be throttled. A throttled expansion occurs at constant enthalpy and is categorized as a Joule–Thomson expansion. If the temperature T_A before expansion is below the inversion temperature then on expansion a decrease in temperature (to T_E Fig. 5.13b) will occur. Assume the cylinder head becomes perfectly conducting and the fluid is heated to the final condition, T_B, by heat transfer through the cylinder head from the surroundings. The heat abstracted is represented by the area E–B–C–D on Fig. 5.13b which is less than area A–B–C–D on Fig. 5.13a. Thus for gas experiencing Joule–Thompson expansion the refrigerating effect can be as little as one-third of that with isothermal expansion even though the fluid started and finished at the same states. In the isenthalpic process no work was done; the piston remained stationary at the lower dead point and the cylinder volume was constant.

For the isentropic process, Fig. 5.13c assume the same sequence as the isothermal case. The cylinder is filled with gas at state A and expanded to a low pressure. The only difference is that the cylinder head is perfectly insulated during expansion and no heat is transferred during expansion so the entropy remains constant ($S = Q/T = 0$). Work is done on the piston and the temperature is reduced as shown by the process A–E in Fig. 5.13c. The cylinder head then becomes perfectly conducting and heat is transferred from the cylinder surroundings to heat the gas to the final temperature, T_B. The shaded area E–B–C–D in Fig. 5.13c is larger than the isenthalpic process and smaller than the isothermal process.

The polytropic process of Fig. 5.13d follows a similar pattern to the isentropic with the difference that the temperature at the end of the expansion T_E' is greater than at the end of isentropic expansion, T_E. This difference may arise from two causes, limited heat transfer during the expansion (usually the case with reciprocating engines but not turbines) and irreversibilities (principally friction) which occur during expansion (particularly in turbines) to increase entropy and, hence, the fluid temperatures. The effect of limited heat transfer and irreversibilities during expansion is to decrease the potential refrigerating effect compared with the isentropic process. The potential refrigerating effect is the shaded area E′–B′–C′–D in Fig. 5.13d smaller than the area for the isentropic process by area E–E′–C′–C.

Fig. 5.13 shows clearly that in terms of refrigerating effect the order of preference of the four processes is (1) isothermal, (2) isentropic, (3) polytropic, and (4) isenthalpic.

In practice isothermal expansion cannot be achieved. It would require infinite rates of heat transfer or the expansion to proceed infinitely slowly. It is therefore customary to adopt an isentropic process as the reference for expansion engines. This too cannot be achieved in practice and so realistic cycle analyses adopt an arbitrary value (60–85 per cent) for the *isentropic efficiency* to account for the polytropic process. The isentropic efficiency is simply the ratio of the actual work of the expansion compared with the work of isentropic expansion.

The same work is necessary to compress the gas regardless of the subsequent expansion. Thus, since the Joule–Thomson process produces so little refrigerating effect compared with isentropic expansion it is to be expected that Joule–Thomson cryocoolers will be less efficient than those incorporating expansion engines. This expectation is realized in practice.

The principal reason for using Joule–Thomson expansion is simplicity. It is much easier and cheaper to provide a simple throttle valve than an expansion engine with means to absorb the work produced. Joule–Thomson expansion is a very common process. All small domestic refrigeration systems operating on the vapour-compression cycle have Joule–Thomson expansion valves rather than small expansion engines.

In larger refrigerating and cryogenic cooling systems the additional expense of expansion engines can be justified by the improved efficiencies. Even in these systems Joule–Thomson expansions are commonly used for the final stage of expansion when there is partial liquefaction of the working fluid because two-phase working fluids are difficult to handle in expansion engines, particularly turbo-expanders.

Daunt (1956) discussed isenthalpic and isentropic expansion in more formal thermodynamic terms than the intuitive-descriptive method followed above. The following is abstracted from Daunt's presentation: 'Writing the first and second laws of thermodynamics as:

$$T\,\mathrm{d}S = C_\mathrm{p}\,\mathrm{d}T - T\left(\frac{\partial v}{\partial T}\right)_p \mathrm{d}p \tag{5.12}$$

we obtain for the case of isentropic expansion ($\mathrm{d}S = \mathrm{O}$) an immediate evaluation for the coefficient of cooling,

$$\alpha_s = \left(\frac{\partial T}{\partial p}\right)_s = \frac{T}{C_{\mathrm{p}}}\left(\frac{\partial v}{\partial T}\right)_p . \tag{5.13}$$

This compares with the Joule–Thomson coefficient of cooling:

$$\alpha_H = \left(\frac{\partial T}{\partial p}\right)_H = \frac{T}{C_{\mathrm{p}}}\left(\frac{\partial v}{\partial T}\right)_p - \frac{v}{C_{\mathrm{p}}} \tag{5.14}$$

whence

$$\alpha_H = \alpha_s - \frac{v}{C_{\mathrm{p}}} \tag{5.15}$$

As p is increased or T decreased, thereby diminishing v, the Joule–Thomson coefficient, α_H, approaches the reversible isentropic cooling coefficient α_s. Moreover as the critical temperature is approached C_{p} tends to an infinite value; hence in regions near critical, isenthalpic expansion also must be highly efficient.'

The favorable environment for Joule–Thomson expansion near the critical point is evident on reference to a temperature–entropy diagram. Figure 5.14, a T–S plane for helium, shows two constant-pressure lines, the liquid–vapour envelope, the critical point C, and three isenthalpic curves A–B, D–E, and F–G. The isenthalpe A–B, starting at the lowest initial temperature, T_{A}, has a high slope not too significantly different from the isentrope A–H. At higher temperatures expansion over the same pressure range results in a much smaller temperature decrease (isenthalpe D–E). If the initial temperature, T_{F}, is above the inversion curve the temperature will actually increase on isenthalpic expansion.

The relationship between α_H and α_s can be written (from Eqs 5.3) and (5.14) as

$$\alpha_H/\alpha_s = 1 - \frac{v}{T}\left(\frac{\partial T}{\partial v}\right)_p . \tag{5.16}$$

This relationship for air as a function of pressure for different values of temperature T (start of expansion) is given in Fig. 5.15 (after Daunt 1956). As the temperature reduces the ratio (α_H/α_s) increases and at 100 K approaches unity. At higher temperatures the ratio (α_H/α_S) is a strong function of pressure, increasing progressively with increase in pressure.

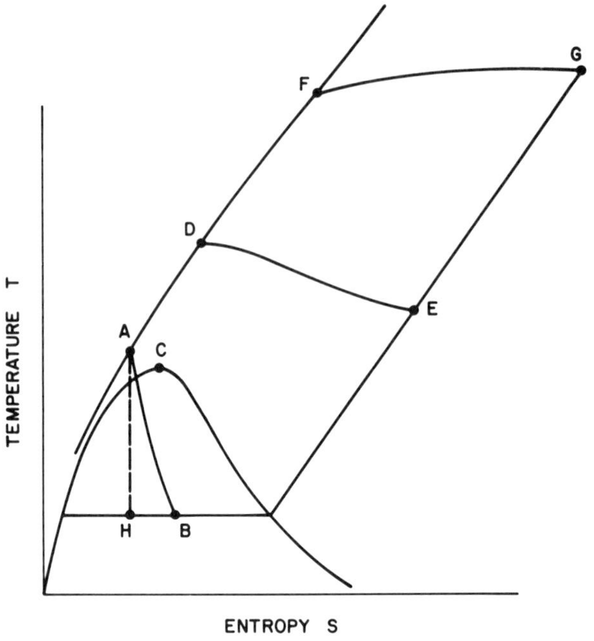

Fig. 5.14 Comparative slope of isenthalpes on the T–S plane.

Joule–Thomson expanders Joule–Thomson expansion nozzles were discussed at length in Chapter 3. Here we limit discussion to say briefly there are many ways to accomplish the isenthalpic expansion from a high pressure to a low pressure albeit inefficient thermodynamically. In commercial service Joule–Thomson expansion nozzles may be broadly classified in two types: (a) fixed-orifice nozzles and (b) variable-flow or demand flow nozzles. Demand flow nozzles are more complicated but orders of magnitude better in performance than fixed-orifice nozzles (for more see Chapter 3).

Expansion engines Expansion engines may be classified in two principal categories:

(1) positive-displacement expanders, usually reciprocating engines;

(2) rotary turbo-expanders.

Reciprocating expanders are typically slow-speed, low mass flow devices able to accommodate virtually any pressure ratio, either in single-stage or multiple expansion engines. In practice, multiple-stage expansion

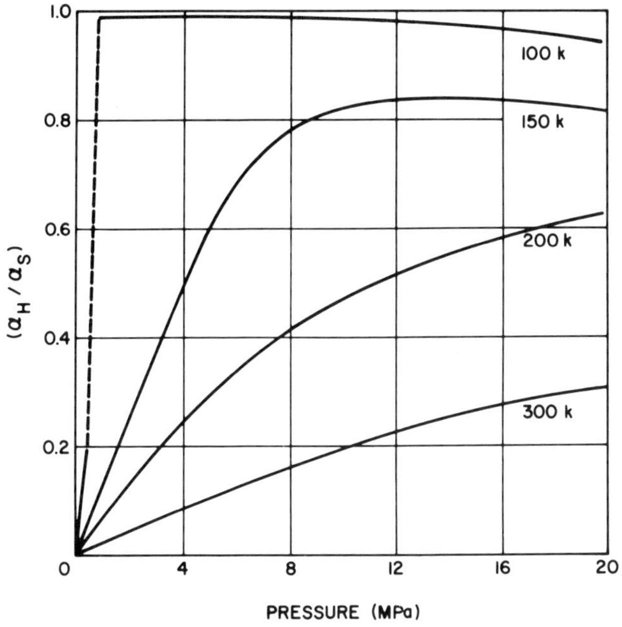

Fig. 5.15 Ratio of the isenthalpic/isentropic coefficients of cooling for air as a function of the initial pressure and temperature (after Daunt 1956).

engines are used in some cryocoolers (such as the Collins type and the Claude cycle helium liquefier) but in a special way. The several identical engines are operated over the same pressure range (i.e. in parallel) but at different temperature levels. This is, of course, contrary to the normal use of multistage expansion steam or air engines. In these applications the various stages operate in series with the pressure limits for the various cylinders decreasing progressively as the expansion proceeds and the cylinder diameters increasing as the fluid density decreases.

Figure 5.11 shows a temperature–entropy diagram for an expansion engine expanding gas from a high pressure p_{max} to a low pressure p_{min}. Several different expansion processes are shown. From the initial high pressure the fluid may expand:

(1) isothermally, process (1–2), represented by $pV = \text{const}$;

(2) isentropically, process (1–4), represented by $pV^{\gamma} = \text{const}$, where γ is the ratio of specific heat at constant pressure and constant temperature ($\gamma = C_p/C_v$);

(3) polytropically, process (1–3), represented by $pV_n = \text{const}$, where $1 \leqslant n \leqslant \gamma$;

(4) polytropically, process (1–5), represented by the equation $pV^m = \text{const}$, where $m \geqslant \gamma$.

In practice the process of expansion is never isothermal or isentropic. In reciprocating machines it lies somewhere between the two (process (1–3)). It cannot be isothermal for this would require infinite rates of heat transfer between the gas and its surrounds. Similarly it cannot be isentropic for this would require perfect insulation. In a turbine, aerodynamic friction causes a higher exit temperature, T_3 than with isentropic expansion because the frictional degradation causes heating of the fluid during expansion.

The expansion process (1–5) cannot normally be attained for it requires the withdrawal of heat during expansion with consequent reduction of entropy. It would only occur in an expander equipped to refrigerate the gas during the expansion process. It is thus of little interest, for our main purpose is to use the expansion process to generate refrigeration rather than consume it.

In reciprocating expanders point 1 on the $P-V$ diagram, Fig. 5.11, is called the *cut-off point*. It is the point where admission of working fluid is terminated and expansion begins. After expansion the exhaust valve opens and fluid is expelled from the cylinder.

The complete cycle for an expansion engine is the $P-V$ diagram 1–2–8–9. The area of the $P-V$ diagram is the work output of the engine. The refrigeration generated is:

(1) $\mathrm{d}Q = \mathrm{d}U + \mathrm{d}W$ for a non-flow system (reciprocating expander);

(2) $\mathrm{d}Q = \mathrm{d}H$ for a flow system (turbo–expander).

To increase the refrigeration the cut-off may be delayed to 6. The fluid expands to the end of the piston stroke at 7, the exhaust valve opens, and the fluid 'blows down' to low pressure 2. Delay in the cut-off point allows more fluid to enter the cylinder and increases the size of the $P-V$ work diagram and hence the refrigeration effect but at the expense of increased use of compressed gas. The process of free expansion (7–2) in the exhaust blow-down occurs as a Joule–Thomson expansion (constant enthalpy) so the fluid is used less effectively than by controlled expansion in the engine. Thus, with delayed cut-off the refrigeration produced by the engine is increased even though the refrigeration per unit mass flow is reduced.

An expander with variable cut-off can be used where a large refrigeration capacity is required for starting and cool-down purposes. Later the cut-off point can be adjusted to provide maintenance levels of refrigeration.

Reciprocating expanders Reciprocating expansion engines in cryocoolers present challenging and unusual design problems. These arise principally from the very low temperatures at which the units are required to operate. Firstly the metals used for the cold parts must be chosen from those which do not exhibit the ductile–brittle transformation suffered by many iron/carbon metals customarily used for expansion machines. Metals suitable for use include stainless steel, the high-nickel alloys, aluminum, and copper. Various surface treatments (nitriding, anodizing, electroplating, and so on) are available to produce glass-hard surfaces if necessary.

During start-up when the expansion engine cools from ambient temperature to the low operating temperature the cold parts will experience some contraction, depending on the coefficient of thermal contraction of the materials used. It is good practice therefore to select the materials for pistons and cylinders that have compatible coefficients of thermal expansion. The alternative is to design the piston/cylinder combination to have the proper running clearance at the normal operating temperature with, perhaps, excessive clearance at ambient temperatures. This requires that a material of low coefficient of thermal contraction be used for the piston and a material with the higher coefficient be used for the cylinder. In cooling the cylinder will contract more than the piston therefore decreasing the clearance.

Contrarily, if the material with the higher coefficient of thermal contraction is used for the piston and, if the piston/cylinder are a well-matched close-tolerance pair at ambient temperature, then, on cooling, the piston will contract more and cause a large high-leakage clearance in the cylinder at normal operating temperatures.

Identical materials may of course be used for both the piston and cylinder and the pair sized to a close tolerance at ambient temperatures. This should also be satisfactory for operation at the cold temperature but great care must be exercised during the cool-down process. The piston is frequently relatively massive compared with the light wall of the cylinder and so cools at a rate slower than the cylinder. This can result in catastrophic seizure of the two parts during the cool-down process because the lighter cylinder wall cools faster and contracts on to the massive piston.

Frictional rubbing and other heat-generating processes must be avoided at all costs in the cold expansion cylinder. Any heat generated is

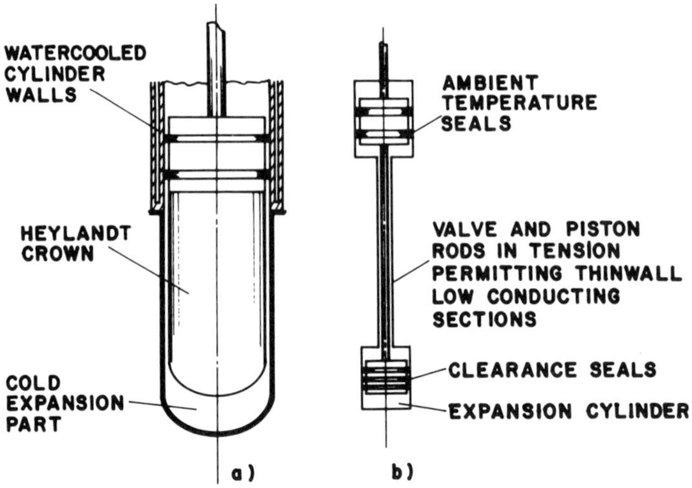

Fig. 5.16 Isolating seals and piston rings in cold expansion engines.

immediately deducted from the 'bottom line' of available refrigeration. Thus, rubbing seals and piston rings should be located remote from the cold region. Design approaches to accomplish this are illustrated in Fig. 5.16.

In one case an extension of the piston, called a Heylandt crown, effectively separates the cold region from the piston carrying the sealing and guide rings. These may operate on water-cooled walls and because they are at ambient temperature the cylinder or liner may be made of the customary graphited cast iron favoured for cylinders in other reciprocating machines.

Figure 5.16(b) shows a piston rod seal operating at normal temperatures at the upper end of a long piston rod with the cold region at the bottom. In most applications it is advantageous to arrange the piston rod and valve rods to be in tension rather than the customary compression. This permits the use of very low cross-section to minimize the thermal conduction heat leaks. Long thin compression members (columns) fail by buckling at loads only a fraction of those that can easily be carried by long thin tensile members, such as multistrand helical wound wire cable.

Rubbing seals on reciprocating pistons in the cold cylinder are best avoided in favour of close-tolerance seals. Parts should be fabricated from materials which provide the possibility for hardened mating surfaces that may be finish ground then lapped to the desired clearance, typically 0.0001 cm per cm of piston diameter, and sometimes better, if the craft skill to accomplish this is available.

With such tight manufacturing tolerances it is vital to avoid designs which involve multiple bores of high concentricity and alignment. Often the manufacturing and operating difficulties can be dramatically eased with the introduction of an articulated member with the limited freedom to radially self-adjust to the optimum degree of concentricity and alignment.

Turbo-expanders Turbo-expanders, like turbo-compressors, are characteristically high mass flow devices. In a turbo-expander the fluid expands from a high pressure to a low pressure in a near-isentropic process with the production of some work. This is accomplished by first expanding the fluid in a nozzle to convert pressure energy to kinetic energy, manifest in a high-speed jet of fluid. The high-speed fluid is then directed into the turbine blades on the rotating disc or wheel. The blades guide the high-speed jet to change its direction and this changes the velocity of the jet and hence the momentum (velocity involves both speed and direction). It is this change in momentum that results in a force applied to the blades, turning the turbine wheel against the resistance of an applied load thus producing some work.

The turbo-expander is thus akin to a reverse image of the turbo-compressor. This compressor absorbs work to accelerate fluid (increase the kinetic energy) which is subsequently converted to pressure energy in the diffuser. The big difference between turbo-expanders and turbo-compressors is that, in the expander, the fluid is 'falling downhill' from a high pressure to a low pressure, a process that will proceed naturally almost regardless of the quality of engineering design of the nozzle and turbine blades. High expansion ratios can therefore be accomplished in a few stages.

In the turbo-compressor, on the other hand, the fluid must be 'coaxed uphill' from a low pressure to a high pressure against its natural inclination. This can only be accomplished in small steps by very careful design and execution of the compressor blades and diffuser section. Many stages of compression are required (for example, 12:1) for every stage of turbine expander handling fluid over the same overall pressure ratio.

Turbo systems are invariably used for the large-scale production of industrial cryogenic liquids and gas separation plants, such as tonnage oxygen or liquid natural gas plants. They are also attractive for cryo-coolers of smaller capacity. The principal advantage is that the rotating assemblies are symmetric and so are perfectly balanced for operation at high speed without rubbing seals and with main shaft bearings that are either oil lubricated or gas lubricated. Gas lubrication using the process gas as a lubricant is particularly attractive to eliminate the need for seals to separate the lubricant and the process gas and to avoid the possibility of contamination.

In some cryogenic plants of intermediate capacity it is customary to combine a positive-displacement compressor, such as perhaps a high-speed, oil-flooded, helical-screw compressor, with a low-temperature high-speed turbo-expander. This is an elegant solution that avoids the relatively high maintenance problems characteristic of reciprocating machinery, combines the compact high mass flow characteristics of rotating equipment and yet avoids the problem of the high cost and relative complexity of the many stages of compression needed for the turbo-compressor to produce an appropriate compression ratio, such as 5 or 8 to 1.

Persistent attempts have been made to apply turbo-expanders and compressors in miniature cryocoolers of very low capacity. Early efforts at the General Electric Company, Schenactady, were described by Gessner and Colyer (1968). More recent work has been reviewed by Valenzuela *et al.* (1987).

In miniature systems the mass rate of fluid flow is very low so the rotating parts are very small. For proper functioning of the expander it is necessary that the turbine blade speed is close to the theoretical optimum fraction of the blade speed/gas speed ratio. The expansion ratio of the gas establishes the velocity of the gas leaving the expansion nozzle. With very small diameter turbine wheels the speed of rotation must be very high since the blade speed $V_b = \pi dn$ where V_b is the mean blade speed, d is the mean diameter of the turbine blades, and n is the speed of rotation.

Consequently in miniature turbines it is not uncommon to find shaft speeds of 500,000 to 800,000 revolutions per minute (approximately 110 MHz), speeds that can only be contemplated with gas bearings. Much of the exploratory development work in this field was accomplished by Herbert Sixsmith originally at the National Bureau of Standards, Boulder, Colorado, the University of Oxford and, more recently, at Creare Inc., Hanover, New Hampshire (Sixsmith *et al.* 1990).

It is generally accepted that the lower limit of capacity of these miniature turbo systems is 5–10 watts at 80 K with the possibility of extension to 1 watt at 10 K. No information about the feasibility and cost of such miniature turbo systems for relatively large-scale series production appears to have been published.

5.4.3 Heat exchangers

Contraflow recuperative heat exchangers are critical elements of all mechanical cryocoolers whether or not they include regenerative heat exchangers. Water- or air-cooled heat exchangers are required to cool the compressed gas between stages (intercooler) of a multiple-stage gas compressor and on leaving the compressor (aftercooler). Further heat exchangers of the contraflow recuperative type are required to cool the compressed gas *en route* to the Joule–Thomson expansion valve or to the expansion engine where the cryogenic temperature will be developed. The low-pressure gas returning to the compressor passes also through the heat exchanger. It is warmed to near ambient temperature by heat transfer from the compressed gas being cooled by the returning gas.

There is a very wide range of different types of heat exchanger, surfaces, materials, and methods of construction.

The specific application of heat exchangers to cryocoolers has been addressed in some detail by Walker (1983). Heat exchangers are reviewed in a more general way by Walker (1990). Design data and design methods for compact heat exchangers of the type most suitable for cryocoolers was given by Kays and London (1976) and by Shah and London (1982).

Many of the papers on cryocoolers in the proceedings of the International Cryocooler Conferences, the Cryogenic Engineering Conference, and the International Cryogenic Engineering Conference contain the details, albeit brief, on many novel design approaches to compact heat exchangers for low-capacity cryocoolers.

5.5 Commercial applications of non-regenerative low-capacity cryocoolers

Joule–Brayton and Claude cycle cryocoolers are widely used for the production of cryogenic liquids, the separation of gases, and the processing of gas mixtures. Large turbo-expanders and compressors are used with large complicated heat exchanger control systems and other features to enhance performance. These large systems are beyond the scope of this volume; our interest is focused on low-capacity cryogenic refrigeration, however imprecisely defined.

One of the most significant developments of low-capacity cryogenic cooling systems occurred in the late 1940s when Collins (1947) made a multiple-expansion Claude cycle system capable of liquefying helium. Claude cycle helium liquefiers had, of course, been constructed earlier. The first to use a reciprocating engine was described by Kapitza (1934). This was a single-stage precooled Claude cycle system. The Kapitza engine was further notable as the first to use a hydrodynamic gas-lubricated piston to minimize frictional heat generation at low temperature.

Collins incorporated some of the ideas exemplified in the Kapitza engine and introduced other innovative features including the use of

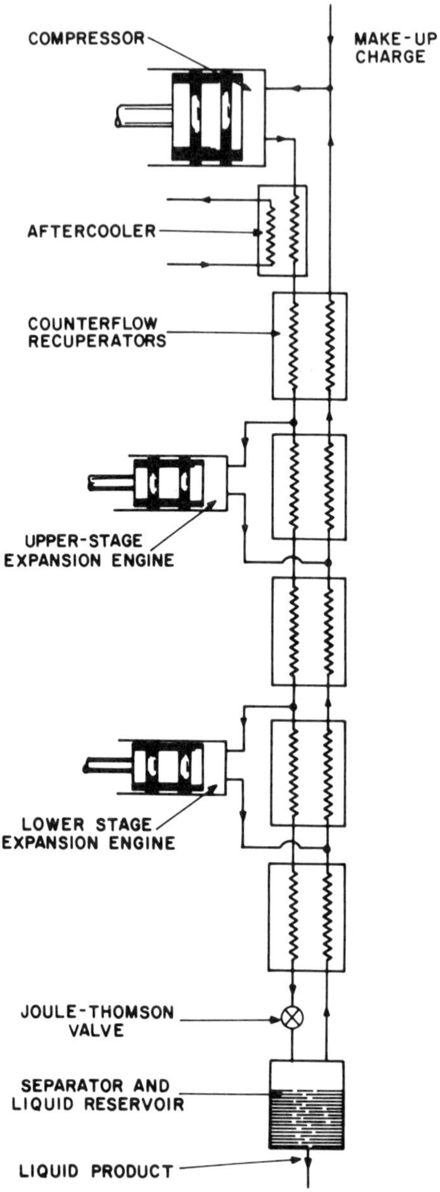

Fig. 5.17 Schematic diagram of the Collins–Claude cycle multiple-expansion gas liquefier.

multiple-expansion engines. The system developed was capable of liquefying any gas (air, hydrogen, or helium) without the benefit of precooling although of course performance was always improved by precooling.

A schematic diagram of the Collins helium liquefier is shown in Fig. 5.17. For the sake of simplicity only two engines are shown, but, depending on the helium inlet pressure, systems with up to five expansion engines have been used. The expansion engines can be physically separated, but coupled to a common crankshaft or can be the stepped (differential) piston type. A Joule–Thomson valve is usually incorporated as the final stage of expansion in the liquefier flow circuit but wet expansion engines with the liquid helium produced in the cylinder have also been used.

Typical performance details for a two-stage Collins liquefier were given by Weinstock (1968). Approximately 25 per cent of the helium flow is expanded in the first expansion engine to provide cooling in the 30–60 K range. About half the helium flow is expanded in the second expansion engine to provide cooling in the 8–15 K range. The remainder is expanded in the Joule–Thomson valve. Liquid nitrogen precooling of the helium, before expansion in the engines, increases the fraction of helium liquefied by two or even three times the production with no precooling.

The Collins liquefier became the virtual standard machine for use in the Western world. In a technical memorandum (CTI 1978) the manufacturers claimed that over 125 of the smaller model 1410 liquefiers were in use in over 20 countries. A decade earlier Weinstock (1968), quoting a private communication, indicated that 500 Collins liquefiers were sold by A. D. Little in the twenty years since the prototype development in 1946. Whatever the number it is certainly true that up to 1950 the number of laboratories dealing with liquid helium was limited to a handful of specialist operations. The development of the Collins machine put the production of liquid helium on to a routine push-button basis and thereby transformed the capabilities and potential of low-temperature research.

Collins evolved the design of the helium liquefier through a variety of systems. In their book Collins and Cannaday (1958) describe in detail an early diaphragm engine, completed in 1938, capable of attaining temperatures of 10 K. The engine experienced minor leakage problems which had a highly deleterious effect on the vacuum insulation surrounding the engine. The Second World War intervened and Collins concentrated his attention on the development of the low-pressure air liquefier with reversing recuperative heat exchangers used for oxygen production. The system incorporated expansion engines for relatively low inlet pressure of 0.6–1.5 MPa.

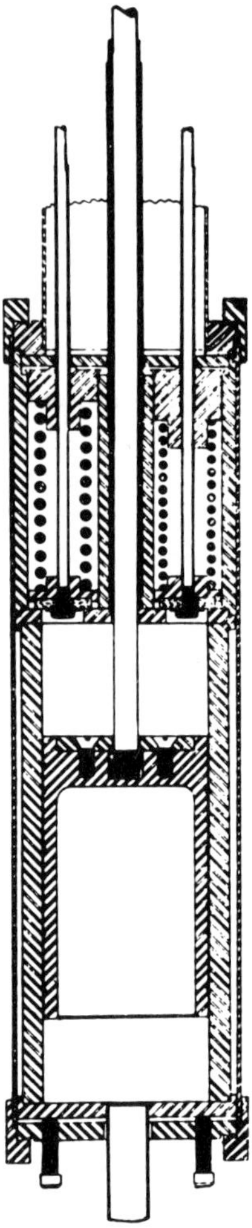

Fig. 5.18 Section of a low-pressure Collins expansion engine.

After the war Collins did not resume work on the diaphragm engines but incorporated his wartime experience in a multiple-expansion helium liquefier using reciprocating expansion engines. Figure 5.18 is a section view of a Collins expansion engine cylinder and valve chamber. Particular features that led to successful operation include the use of a very close fitting piston (of soft material over a steel core) sliding in a cylinder with hardened surface, a highly efficient continuous recuperative heat exchanger, and the design innovation of putting the piston and valve rods in tension rather than the customary compression mode. This latter feature permitted the use of thin flexible rods reducing the conductive heat leak and facilitating free movement of the piston in the cylinder.

Another important feature was to contain the engines and heat exchangers in a stainless steel dewar vessel containing the liquid helium product. The external environment of the engines and heat exchangers was low-pressure helium, therefore minor leaks from the pressurized engine/recuperator system were of no significance compared with a similar leak to a vacuum enclosure.

Further construction and operating details of the system were given by Collins (1947, 1952, 1956, and 1966) and Collins and Cannaday (1958). The following is a description of the chief characteristics of the helium liquefier by Collins and Cannaday (1958):

Rate of liquefaction	25–32 l/hr
Power required	45 kW
Helium circulated	215 g mol/min (185 cfm)
Operating pressure	12.5 atm
Refrigeration	Expansion engine plus liquid nitrogen
Liquid nitrogen	(If no liquid nitrogen is used, rate of helium liquefaction is 10 l/hr)
Heat exchanger	Hampson type
Actual work expended (N_2 plant included)	3.1 kW hr/l
Computed requirement (actual liquefier but with N_2 plant and helium compressor assumed reversible)	0.87 kW hr/l
Computed requirement (entire process reversible)	0.24 kW hr/l

5.5.1 *Description of the liquefier (Collins and Cannaday 1958)*

A flow diagram of the actual cycle is given in Fig. 5.19. It differs from the idealized cycle, not only because the heat exchangers and expansion engines are necessarily imperfect, but also because practical considerations have influenced the choice of apparatus and procedures. For the sake of compactness of the liquefier and greater production of liquid from variable compressed helium, liquid nitrogen is employed to the

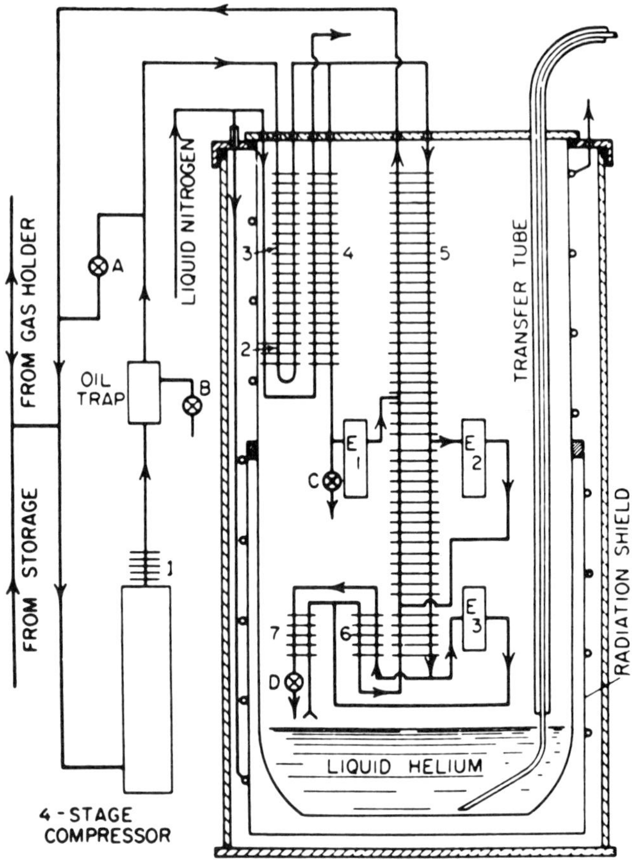

Fig. 5.19 Flow diagram of Collins helium liquefier.

extent of its utility. The transfer of heat from gaseous helium to liquid nitrogen evaporating at a constant temperature is irreversible, of course, and a net increase of entropy is incurred. A final difference lies in the substitution of a throttle valve for the fourth engine of the cycle.

In the flow diagram shown (Fig. 5.19), compressed helium (about 12 atm) from the compressor is treated for entrained oil in an oil trap and for vaporized oil in a refrigerated heat exchanger 3. Thereafter the stream divides, about 8 per cent going to heat exchanger 4, in which it is cooled to 80 K by means of liquid nitrogen and then expanded in engine E_1, the remainder going to the principal heat exchanger 5. The temperature of the gas in heat exchanger 5 ranges from room temperature at the upper end to 15 K at the bottom. At the zone of exchanger 5 where the temperature is 40–45 K, a second fraction (about 15 per cent of the whole) of the compressed helium is led off for expansion in the second engine, E_2. At the lower end of exchanger 5 a final division of the stream occurs. About 52 per cent is used in expansion engine E_2 and 25 per cent flows through the small exchangers 6 and 7 to the expansion valve D.

All the helium that enters exchanger 4 passes through engine E_1, the pressure falling to substantially one atmosphere and the temperature falling from 80 K to about 45 K The expanded helium joins the lower-pressure stream in exchanger 5. All three engines are served by a single crankshaft and are identical in size. The bore is 2 inches, and the stroke is 2 inches. Each engine is comprised of a single cylinder. Although the three engines receive compressed helium from the same supply line and discharge their spent gas into the same low-pressure conduit, they operate at different temperature levels and embrace different sections of the heat-exchange system. Helium enters E_2 at about 45 K and is discharged at about 25 K The inlet temperature of E_3 is about 15 K and the outlet temperature is estimated to be about 9 K.

Liquid helium forming at the throttle valve D drops into the bottom of the dewar vessel in which the heat exchangers and engines are suspended. Space for 30 litres is provided.

Certain features of the earlier helium cryostat have been retained. The heat exchangers and engines, which hang from a steel plate, are surrounded by an atmosphere of helium rather than by the insulating vacuum. Minor leaks from the high-pressure stream can be tolerated. The helium atmosphere is contained by a large metal dewar vessel, the vacuum jacket of which is continuously pumped. The lower half of the inner wall of the vacuum jacket is enclosed by a nitrogen-cooled radiation shield.

The engine cylinders and pistons are made of nitrided nitralloy and are so closely fitted that piston rings are unnecessary. The piston rods are relatively long and slender, and are made to operate in tension to promote perfection of alignment of the piston within the cylinder. There exists a thermal gradient in the helium atmosphere surrounding the heat exchangers and engines, the temperature being approximately 295 K at the top and 4.2 K at the bottom. As far as practicable, the cylinders of the engines are located at the proper elevation for matching temperatures inside with outside in order to reduce convection to a minimum. The stuffing boxes for the piston rods and valve pull rods and the running gear of the engines are placed on top of the lid of the dewar so that heat generated in these parts can be kept out of the cold region.

The piston rods and valve pull rods are attached to the ends of horizontal walking beams, as in ancient steam engines. The beams are

Fig. 5.20 Collins–Claude cycle helium liquefier (CTI Model 1430).

2 feet in length, are pivoted at one end, and at the point immediately above the crankshaft are fitted with ball bearings to act as cam followers. With a stroke of only 2 inches the end of the piston rod travelling in an arc of 2-foot radius is not pulled away from the vertical by an appreciable angle. Speed control is achieved by centrifugal action of a split flywheel within a brake drum. The two halves of the flywheel are fitted with brake shoes. Speed is adjusted by changing the compression of a spring. This adjustment may be made while the engine is running.

The Collins engine became the prototype for a family of helium liquefier systems in series production up to the present time. Figure 5.20 shows the model 1430 helium liquefier system formerly in production at CTI, Waltham, Massachusetts, and now manufactured by Process Systems International. The system has the capacity to produce 5–40 litres per hour of liquid helium or 19–100 W of refrigeration at 4.6 K depending on the compressor option and mode of operation chosen. For maximum output the helium refrigerator module requires 40 litres per hour of liquid nitrogen and 2.5 kW of electric power. The liquefier module weighs 770 kg and is contained in an envelope 1.27 × 30.375 × 0.69 m.

The associated helium compressor module requires 100 kW of electric power and 70 kg per minute of cooling water at 24 °C and 0.3 MPa. The mass of the compressor module is 1455 kg and is contained in an envelope 0.415 × 0.475 × 0.715 m.

Following the pioneering work by Collins at the Massachusetts Institute of Technology and the initial development of the commercial version at A. D. Little, Cambridge, Massachusetts, successive improvements and variants appeared. These were reviewed by Walker (1983), but none have gone into series production on the same scale as the Collins machine.

The Collins–Claude liquefier in various forms continued in production at CTI, Waltham, Massachusetts for many years until, about 1980, the manufacture and distribution were assumed by Process Systems International, Bedford, Massachusetts.

Installation of a Collins–Claude helium liquefier represents a significant investment ($100,000 or more in 1989). Moreover, the liquefier has a fairly high maintenance requirement, not enough to occupy a skilled technician full-time but enough to require that such a person be available on a frequent and regular basis.

The increasing use of superconducting magnet systems (nuclear magnetic resonance imaging systems and the like) requiring cooling with liquid helium has spurred renewed interest in the possibilities for a low-capacity (10 litres per day) helium liquefier having a low or zero maintenance requirement apart from say annual replacement of a rebuilt unit. Various applications of two-stage Gifford–McMahon cryocoolers with a supplementary Joule–Thomson circuit have been described in the literature.

Other approaches to small-scale reliable helium liquefiers are in development. Harada *et al.* (1987) of Hitachi Ltd., Japan have described a prototype Claude cycle unit with a two-stage turbo-expander specifically aimed at the maintenance and vibration-free systems required for NMR systems and other cryogenic electronic (the supercomputer) applications. Walker (1987) described a four-stage expansion Stirling cryocooler with a four-stage expansion Stirling cryocooler with final-stage helium Joule–Thomson expansion in early development at the General Pneumatics Corporation, Scottsdale, Arizona.

5.6 Doll–Eder valveless expansion engines

Two researchers at the Bavarian Academy of Sciences in Munich, West Germany, R. Doll and F. X. Eder (1964) introduced an important new concept in expansion engine technology: the valveless engine with

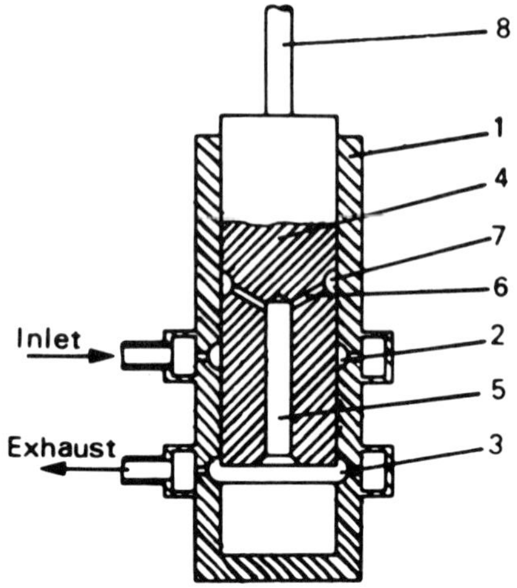

Fig. 5.21 Cross-section of a Doll–Eder valveless expansion engine.

hydrodynamic gas-lubricated piston. A simplified cross-section of the engine is shown in Fig. 5.21 reproduced from Kneuer and Turnwald (1968). The engine cylinder (1) has two annular channels (2) and (3) for the admission of high-pressure gas and the exhaust of low-pressure gas. The piston (4) has a central bore (5) connected via radial channels (6) to a ring groove (7) on the piston surface. Work is transferred via the piston rod to the crank-gear and hence to a dynamometer.

Gas flow and the resultant work diagram for the engine are illustrated in Fig. 5.22 (after Kneuer and Turnwald 1968). At condition (1), with the piston in the bottom dead centre position, the ring groove on the piston is aligned with the inlet annular channel in the cylinder. High-pressure gas enters the engine filling the clearance space, volume V_1, principally the central bore of the piston. As the piston rises the piston ring groove and the cylinder inlet annular channel move out of alignment thereby closing off admission of further fluid at high inlet pressure. The fluid contained in the cylinder expands to a lower pressure and work is produced.

At condition (2) the lower edge of the piston clears the annular exhaust channel in the cylinder releasing expanded fluid to exhaust. The pressure drops rapidly to the exhaust pressure (condition (3)) as the piston ascends to top dead centre. Finally the piston descends closing off the exhaust annular channel and compressing the gas retained in the cylinder to condition (4). Near the piston top dead centre position the inlet port is

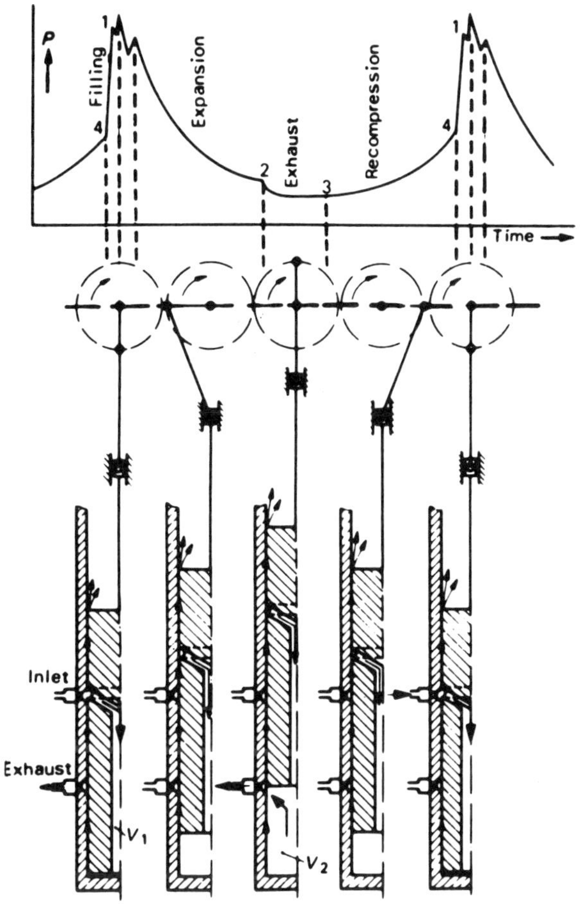

(a)

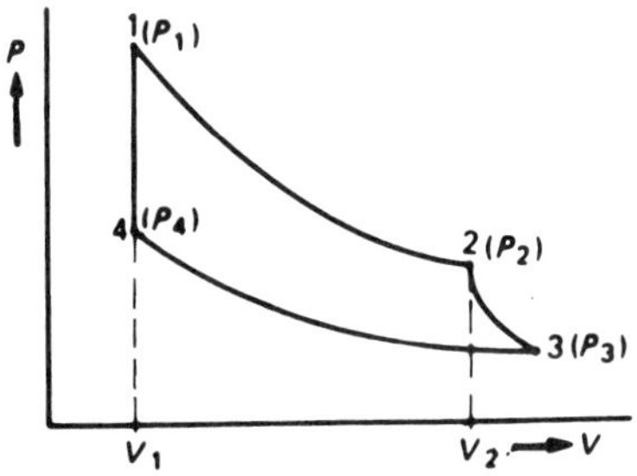

(b)

Fig. 5.22 Operational cycle and work diagram for a Doll–Eder valveless expansion engine.

again uncovered and fresh charge is admitted to the cylinder to repeat the cycle.

There is a leakage of high-pressure gas to exhaust in the annulus formed by the piston in the cylinder. The leakage is indicated, by the arrows on Fig. 5.22, as a flow from (a) the inlet to the exhaust annular channel, (b) the underside of the piston to the exhaust channel, and (c) the inlet channel to the space above the piston. This leakage flow is the hydrodynamic gas lubricant for the piston. Kneuer and Turnwald (1968) identify the flow as a loss that is unavoidable with valveless engines and give the following equation for the magnitude of the flow loss:

$$\dot{v} = \frac{(p_1^2 - p_2^2)\,\alpha^3 U}{24 p_1 L \beta} \tag{5.17}$$

where $\dot{v}$ is the volumetric rate of flow, p_1 is the inlet pressure, p_2 is the exhaust pressure, a is half the clearance of the piston in the cylinder, U is the circumference of the piston, L is the length of the leakage path, and, β is the viscosity of the working fluid.

From this it can be seen that the leakage ratio is a function of the cube of the clearance in the cylinder and compels close-tolerance clearance fitting of the piston in the cylinder. Kneuer and Turnwald (1968) mention a clearance of less than 2 and 3μ with piston diameters of 28 and 38 mm.

Becker *et al.* (1968) in their discussion of the leakage loss suggest that with 2μ clearance the loss is 5–10 per cent of the total gas flow with pressures in the range 0.3–3 MPa. This flow is said to be sufficient to stabilize the piston and provide complete gas lubrication: '... by means of a large number of grooves on the piston surface. The clearance between successive grooves has a proper profile and achieves high stabilizing forces'.

Doll–Eder valveless expansion engines have been used extensively by the German company Linde Aktiengesellschaft of Munich. In 1968 Kneuer and Turnwald wrote that over 50 valveless engines had been incorporated in helium liquefiers in the previous four years. Baldus and Sellmaier (1968) described a complicated system to generate temperatures of 1.8 K in a helium II refrigerator utilizing a Doll–Eder valveless expansion engine in a Claude cycle system with a low-pressure Joule–Thomson expansion circuit generating 30 W of refrigeration at 1.8 K.

Claudet and Verdier (1972) described an interesting prototype expansion engine for Claude cycle systems with valves operated by the piston motion. Figure 5.23 is a sketch of the engine (after Claudet and

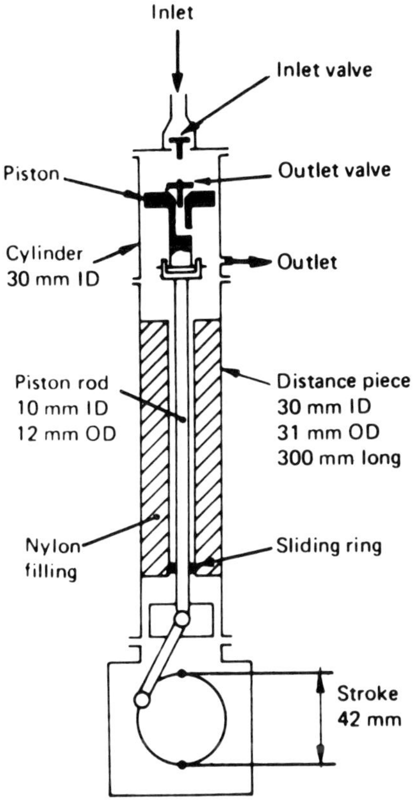

Fig. 5.23 Expansion engine with piston-operated valves.

Verdier 1972) and Fig. 5.24 shows the relative piston and valve positions during operation. The cycle of events in the engine is as follows:

1. *Exhaust*. With the piston at bottom dead centre and beginning its ascent the inlet valve is kept closed by the high pressure of the inlet fluid. The outlet valve located in the piston is opened by a spring F located under the valve. Gas is released from the cylinder to exhaust.

2. *Compression*. Exhaust gas leaves the cylinder as the piston ascends towards top dead centre. As the piston approaches the end of the upward stroke the spring F above the exhaust valve is compressed and closes the exhaust valve. Shortly thereafter the spring F is compressed sufficiently to cause the inlet valve to lift from its seat

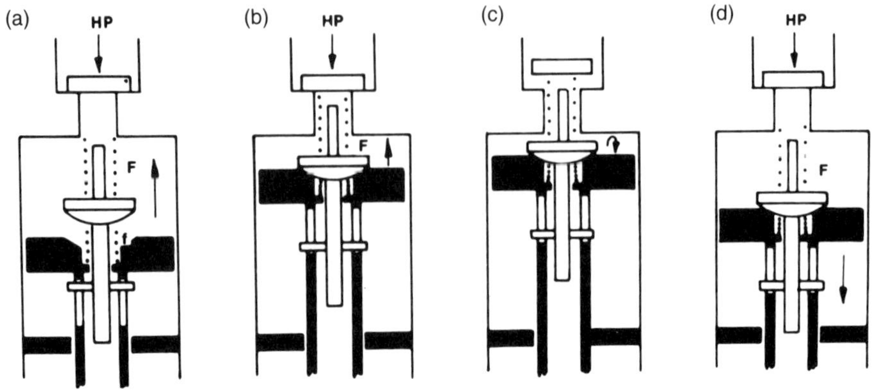

Fig. 5.24 Piston and valve positions in a Claudet–Verdier expansion engine with piston-operated valves.

(alternatively an extension of the exhaust valve impinges on the underside of the inlet valve causing it to lift) and new charge is admitted to the engine.

3. *Induction*. The piston passes the top dead centre position with the inlet valve open and the exhaust valve closed. As the piston descends the spring force F holding the inlet valve open decreases and the inlet valve closes.
4. *Expansion*. The piston descends with both valves closed. The pressure decreases as the expansion proceeds.

The performance of the prototype Claudet–Verdier engine was most encouraging. With inlet–outlet pressures of 0.9 and 0.14 MPa they observed isentropic efficiencies in the range 50–70 per cent. The temperature difference across the engine ranged from:

(1) 5 K with an inlet temperature of 20 K, to

(2) 26 K with an inlet temperature of 76 K.

Caratsch (1991) described an interesting Claude refrigerator of 1 kW capacity at 800 K. The plant consisted of an oil-injected screw compressor compressing nitrogen from 1 to 12 bar at the rate of 35 g/s. The compressed gas was cooled in a water-cooled aftercooler and subsequently in a recuperative exchanger to 123 K. A fraction of the cooled gas (29 g/s) was expanded in two stages of turbo-expander with gas bearings reaching a minimum temperature of 82.5 K. This cold gas stream was used to cool the unexpanded fraction to saturated liquid at 106 K and subcooled to 88 K before Joule–Thomson expansion to 1 bar, 77.6 K

The unit consumed 208 kW of three-phase 60 cycle power to generate 1 kW of refrigeration at 80 K The unit was developed by the Cryogenic Division of Sulzer Bros., Winterthur, Switzerland. The Division has recently (1992) been sold to the German Linde company and the future commercial prospects for this interesting machine are unknown.

5.7 Closed-cycle Joule–Thomson or Linde–Hampson systems

Open-cycle gas storage systems for cryogenic cooling were discussed in Chapter 3. These utilize a Giaque–Hampson heat exchanger for cooling the gas followed by a Joule–Thomson expansion to partially liquefy the gas. In small sizes (fraction of a watt at 80 K) it is often most convenient to energize the system from gas stored in a bottle at very high pressure and allow the exhaust to vent to the atmosphere. Closed-cycle versions of this same system are possible wherein a compressor is included in the circuit. The compressor inhales the exhaust gas to recompress it for subsequent use in the Joule–Thomson expansion. The closed-cycle system is called a Linde–Hampson cryocooler.

A closed-cycle, Linde–Hampson cryocooler made by Allied Signal Garrett AiResearch Division of Torrance, California, is shown in Fig. 5.25. This unit has a capacity of 2.5 watts at 80 K. The cooler uses nitrogen as the working fluid with an electrically driven, three-stage compressor and a post compression, adsorber unit to clean the compressed gas. The compressor operates at high speed, 4700 revolutions per minute, to provide nitrogen at up to 2300 lb per sq in pressure using 28 volt, AC power at 400 Hz. Power consumption is approximately 500 watts. The system weight is 31.5 lb and the envelope dimensions 11.75 in × 16 in × 7.75 in. Two accumulator tanks, 10.85 in × 10.45 in × 5.4 in, are provided on the low-pressure side of the compressor. When the unit is not operating the tanks are pressurized to approximately 60 lb per sq in with an ambient temperature of 75 °F.

The Garrett unit was developed in the 1960s for an aircraft-mounted, infrared mapping system and was later adapted to aircraft-mounted, infrared fire control systems. It is now available in a variety of configurations with over 2000 units manufactured. The scheduled maintenance interval is 100 hours and the mean time between failures is 3800 hours.

Another closed-cycle Linde–Hampson cryocooler is shown in Fig. 5.26. This was developed by the General Pneumatics Corporation, Scottsdale,

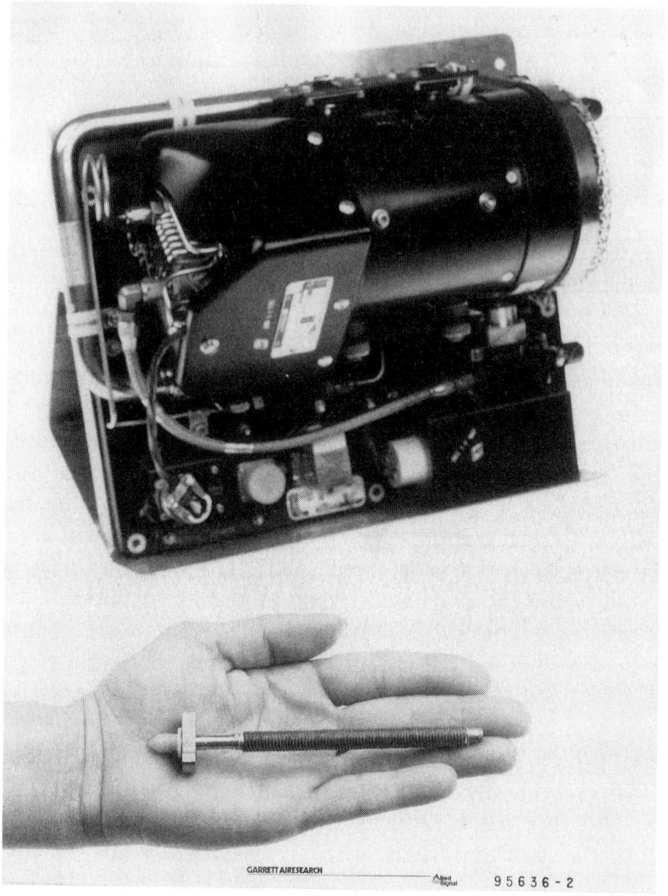

Fig. 5.25 Closed-cycle Linde–Hampson cryocooler (courtesy Allied Signal Garrett AiResearch Div.).

Arizona, for the NASA Kennedy Space Center. It is now in limited production or development for various other purposes including a triple-fluid version which liquefies nitrogen and neon to precool helium before isentropic expansion followed by isenthalpic expansion to 4 K.

The General Pneumatics units include the unusually compact four-stage compressor shown in Fig. 5.27. This was described by Atkins *et al.* (1987). A single reciprocating element is driven by a Scotch-yoke mechanism located approximately at the middle of the longitudinal axis. The single reciprocating element is suspended on two axial stubs protruding from the two end covers. The stubs constitute the third- and fourth-stage pistons mating with the corresponding cylinders in the single

Fig. 5.26 Closed-cycle Linde–Hampson cryocooler (courtesy General Pneumatics Inc.).

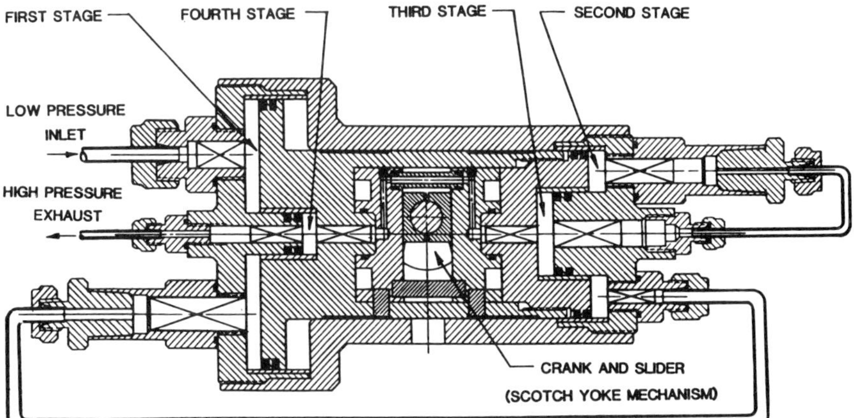

Fig. 5.27 Compact four-stage gas compressor for a closed-cycle Linde–Hampson cryocooler (courtesy General Pneumatics Inc.).

reciprocating element. The first and second compression stages are the end faces of the reciprocating element working in the mating cylinders of the static enclosure.

The initial prototypes of this unit were fitted with dry rubbing piston rings of the glass filled polytetrafluoroethylene material Rulon A. The intent is to eventually equip the unit with close-tolerance gas seals on the first- and second-stage pistons. The third- and fourth-stage pistons will be equipped with close-tolerance gas bearings energized by a fraction of the compressed gas from the third and fourth stages. The gas bearings will be drained to the second- and third-stage cylinders respectively and 'recycled' by recompression.

All the Linde–Hampson systems under development at General Pneumatics combine the compact four-stage gas compressor and the contaminant-tolerant, temperature-sensitive, variable-area flow regulator for Joule–Thomson nozzles discussed earlier in Chapter 3, and shown in Fig. 3.19.

5.8 Closure

In this chapter we have examined, albeit superficially, the technology of closed-cycle mechanical cryocoolers that do not use regenerative heat exchangers. These include machines working on the Linde–Hampson (Joule–Thomson), Claude, and Joule–Brayton cycles. All use compressors and expanders of one sort or another and recuperative heat exchangers, a key component to all systems for cooling the gas prior to expansion.

The non-regenerative systems are widely used for large-scale cryogenic refrigeration but are not so common in smaller sizes as the alternative regenerative systems.

There is a growing need for a reliable low-capacity helium liquefier and this demand is expected to spur the development of a variety of different systems, some of which will surely embody much of the technology discussed above.

References

Atkins, A., Walker, G., and Zylstra, S. (1987). Miniature Multi-Stage High Pressure Gas Compressor for Linde–Hampson Cryocoolers. *Proceedings of the Fourth International Cryocooler Conference* (ed. G. Green *et al.*), David Taylor Naval Ship R. and D. Center, Annapolis, Md.

Baldus, W. and Sellmaier, A. (1968). A Continuous Helium II Refrigerator. *Advances in Cryogenic Engineering*, **18**, 434–40.

Barron, R. (1984). *Cryogenic Systems*, 2nd edn. Monographs on Cryogenics 3, Oxford University Press.

Becker, H., Doll, R., and Eder, F. X. (1968). Measurement of the Efficiency of the Valveless Doll-Eder Expansion Engine. *Proceedings of the Second International Cryogenic Engineering Conference*, Iliffe, Guildford, UK, pp. 9–11.

Caratsch, E. (1991). 1 kW Capacity Refrigeration System at 80 K (Linit-R1). *Proceedings: Applications for Cryogenic Technology*, **10**, Plenum, New York.

Claudet, G. and Verdier, J. (1972). Simplified Cryogenic Reciprocating Expansion Engine. *Proceedings of the Fourth International Cryocooler Conference*, pp. 80–2, Eindhoven. IPC, Guildford, UK.

Collins, S. C. (1947). A Helium Cryostat. *Review of Scientific Instruments*, **18**, 157.

Collins, S. C. (1952). Helium Liquefiers. *Science*, **116**, 289.

Collins, S. C. (1956). Helium Liquefiers and Carriers. In *Handbuch der Physik*, vol. XIV, Springer-Verlag, Berlin. pp. 113–35.

Collins, S. C. (1966). Helium Refrigerator and Liquefier. *Advances in Cryogenic Engineering*, **11**, 11–15.

Collins, S. C. and Cannaday, R. L. (1958). *Expansion Machines for Low Temperature Processes*. Oxford University Press.

Collins, S. C. and Streeter, M. H. (1968). Refrigerator for 1.8 K. *Proceedings of the First Cryogenic Engineering Conference*, Tokyo, pp. 215–17, Heywood-Temple, London.

CTI (1978). Standard Helium Liquefier Model No. 1410. Tech. Mem. 91477, CTI, Waltham, Massachusetts.

Daunt, J. G. (1956). The Production of Low Temperature Down to Hydrogen Temperatures. In *Handbuch der Physik*, vol. IV, Springer-Verlag, Berlin.

Doll, R. and Eder, F. X. (1964). Gas Lubricated Low Temperature Piston Expansion Engine Without Control Valves. *Advances in Cryogenic Engineering*, **9**, 561–4.

Gessner, R. L. and Colyer, D. B. (1968). Miniature Claude and Reverse Brayton Cycle Turbomachinery Refrigerators. *Advances in Cryogenic Engineering*, **13**, 474–84. See also: Colyer, D. B. and Gessner, R. L. (1968). Miniature Cryogenic Refrigerator Turbomachinery. *Advances in Cryogenic Engineering*, **13**, 485–93.

Harada, S., Matsuda, T., Saito, S., and Ihara, K. (1987). Small Size Helium Refrigerator with Micro Turbo Expanders. *Proceedings of the Fourth International Cryocooler Conference* (ed. G. Green *et al.*), David Taylor Naval Ship R. and D. Center, Annapolis, Md.

Johnson, R. W., Collins, S. C., and Smith, J. L. (1971). Hydraulically Operated Two-Phase Helium Expansion Engine. *Advances in Cryogenic Engineering*, **16**, 171–7.

Kapitza, P. (1934). The Liquefaction of Helium by an Adiabatic Method. *Proceeding of the Royal Society*, *London*, Ser. A., **147**, 189–94.

Kays, W. and London, A. L. (1976). *Compact Heat Exchangers*. 2nd edn, McGraw-Hill, New York.

Kneuer, R. and Turnwald, E. (1968). Measurements and Experience with Valveless Piston Expansion Engines. *Proceedings of the Second International*

Cryogenic Engineering Conference, Brighton, UK, pp. 12–15, Iliffe, Guildford, UK.

Shah, R. and London, A. L. (1982). Laminar Flow Heat Transfer. In *Recent Advances in Heat Transfer*, Academic Press, New York.

Sixsmith, H., Hasenbein, R. J and Valenzuela, J. (1990). A Miniature Wet Turboexpander. *Advances in Cryogenic Engineering* (ed. R. Fast), **35**, Plenum, New York.

Valenzuela, J., Sixsmith, H., and Swift, W. (1987). Long Lifetime Closed Cycle Cryocooler for Space. *Proceedings of the Fourth International Cryocooler Conference* (ed. Green *et al.*), David Taylor Naval Ship R. and D. Center, Annapolis, MD.

Walker, G. (1983). *Cryocoolers*. 2 vols. (see Vol. II, Chapter 8, Heat Exchangers in Cryocoolers), Plenum, New York.

Walker, G. (1987). Low Capacity Helium Liquefier: A Ross–Stirling Engine with Zimmerman Displacer and Two-Stage Compression with Joule–Thomson Expansion. *Proceedings of the Fourth International Cryocooler Conference* (ed. G. Green *et al.*), David Taylor Naval Ship R. and D. Center, Annapolis, Md.

Walker, G. (1990). *Industrial Heat Exchangers: The User Guide*. 2nd edn, Hemisphere, Washington, D.C.

Weinstock, H. (1968). A Review of Cryogenic History and Principles of Refrigeration. *Application of Cryogenic Technology*, Vol. 1. Tinnon-Brown, Alhambra, CA.

6

STATIC COOLING SYSTEMS

6.1 Introduction

A miscellany of topics is included in this chapter. They have the common feature that they are static systems or parts of systems, static in the sense of possessing no moving parts, as compared to dynamic systems that do have moving parts.

Long life is intrinsically inherent in the absence of moving parts. This is a compelling advantage for many systems where maintenance and replacement are difficult or impossible. Other principal advantages include the elimination of vibration, noise, mechanical wear, lubricant (contamination of the working fluid), and generation of detritus or wear debris.

One type of system, *radiant coolers*, can achieve temperatures in the range of 80–200 K simply by radiation heat transfer. These coolers require no power supply in addition to having no moving parts. For this reason they are most attractive for spaceborne applications. Unfortunately the refrigerating capacity they can achieve, particularly at the lower temperatures, is severely limited by the practical size and weight of the radiating surfaces that can be provided or tolerated.

Another type of system, *thermoelectric cooling*, operates by virtue of the phenomenon that two dissimilar metals connected in series with an applied electromotive force will be cooled at one junction while heated at the other. Thermoelectric cooling is very reliable and operates independently of orientation or gravity effects. Moreover, the weight and volumes are small, attributes as important as the absence of moving parts in many applications. Unfortunately the materials presently available for thermoelectric cooling do not possess the necessary properties to achieve temperatures in the cryogenic range, even with the complication of multistage units.

Adsorption and absorption compressors (collectively known as *sorption compressors*) are used to compress gases. They operate without moving parts except for inlet and outlet valves which can be made very long lasting. Sorption compressors are of great interest as alternatives to the mechanical compressors used in Linde–Hampson (Joule–Thomson) and Gifford–McMahon cryocoolers. They operate by virtue of the mass of working fluid (gas) absorbed or adsorbed by selected media (activated charcoal, zeolites, and so on) at different temperature levels. Thus by cyclic heating and cooling of the sorption matrix and by synchronized operation of the valves the progressive sorption and release of working fluid with compression to a higher pressure can be achieved. Use of thermal input instead of electrical power has obvious attractions for situations where waste heat is available. Sorption compressors are at an early stage of development and have yet to prove the anticipated capability for long-lived operation with reasonable system weights and volumes.

Magnetic refrigerators utilize the magnetocaloric phenomenon exhibited by some materials, paramagnetic salts, wherein heat is released or absorbed by the imposition and release of a magnetic field in and about the material.

Thus far magnetic refrigerators appear to have been most valuable as the means to achieve very low temperatures, absorbing heat at temperatures of 1 K or less while releasing heat at temperature levels of 10 K or more. They are normally associated with secondary refrigerators operating over the range 10–300 K.

However, magnetic refrigerators operating at higher temperatures up to ambient temperatures have been demonstrated in the laboratory. Engineering designs for high-capacity units at high temperatures have been discussed in the literature. The advantages claimed for magnetic refrigeration include very high efficiency (in terms of the actual performance compared with the theoretical Carnot performance), small size or high capacity per unit volume, and relatively low cost.

The range and variety of systems used or anticipated is very wide. To date no definitive design has emerged and gone into production. At this stage it is not possible to identify the technology 'pinch-points' or to speculate meaningfully on the future of magnetic refrigerators.

6.2 Radiant coolers

A solid surface, whatever its shape, emits thermal radiation according to the equation

$$Q_{\mathrm{E}} = \epsilon A \sigma T \qquad (6.1)$$

where Q_{E} is the rate of radiation heat transfer (watts) emitted by the surface, ϵ is the emissivity, an optical property of the surface ranging from practical limits of 0.04–0.96, A is the external projected area of the surface, σ is the Stefan–Boltzmann radiation constant (5.67×10^{-8} W/m^{2}K^4), and, T is the absolute temperature of the surface.

Loss of heat by radiation heat transfer is the mode of cooling for radiant coolers. The only practical application is on spacecraft where the high vacuum of deep space eliminates the possibility for convective heating. The low effective temperature of deep space (4 K) provides a suitable thermal sink and by careful design it is possible to achieve temperatures in the range 80–200 K with no moving parts and no power input.

6.2.1 *Fundamental considerations*

Radiation energy is emitted from a surface as an energy wave with a characteristic wavelength depending on the temperature. Low-temperature surfaces emit radiation with a long wavelength compared with hotter surfaces emitting short-wave radiation.

The same surface also receives radiation from elsewhere. The surface will either absorb or reflect the radiation incident upon it. The absorptivity α of a surface to incident radiation is an optical property of the surface and ranges from a low value of 0.15 to a high value of 0.95.

The absorptivity α and the emissivity ϵ of a surface depend on the wavelength. If the wavelength of the radiation emitted is approximately the same as the radiation received then $\epsilon = \alpha$ (Kirchoff's relation). Frequently, however, surfaces on a spacecraft receive short-wave, solar radiation (from a very hot body, the sun) while emitting long-wave radiation at relatively low temperatures. The optical properties α_s, the solar absorptivity, and ϵ, the infrared emittance, may be quite different.

In general all surfaces can be classified as:

(1) solar absorbers (primarily polished metals) with a high $\alpha_s = 0.2$ and low $\epsilon = 0.04$;

(2) solar reflectors (specifically developed white paints) with a low $\alpha_s = 0.2$ and high $\epsilon = 0.94$;

(3) flat absorbers (black paint, black copper oxide) with a high $\alpha_s = \epsilon = 0.85$;

(4) flat reflectors (aluminum paint) with moderate-to-low $\alpha_s = \epsilon = 0.2$.

6.2.2 *Some aspects of design*

To be most effective the radiant cooler surface, called the 'cold patch', should clearly be a solar reflector surface emitting infrared (long-wavelength) radiation with the maximum possible emittance. The cold patch must also be protected by screens and reflective shields from receiving radiation inputs from direct or reflected solar radiation (short-wave)

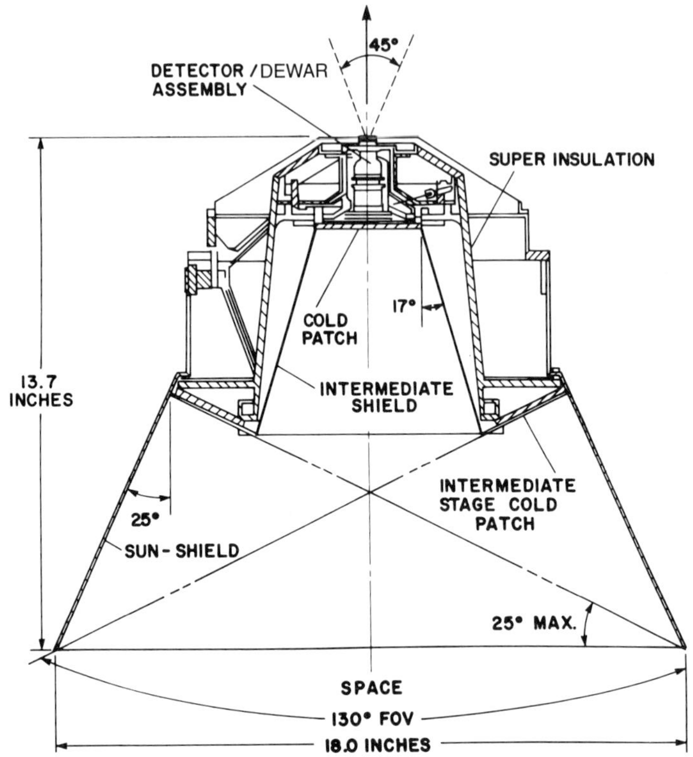

Fig. 6.1 Radiative cryogenic cooler (capacity 2 mW at 80 K) (after Donabedian 1978).

and direct or reflected infrared radiation (long-wavelength) radiation from the earth, or from other surfaces on the spacecraft, including, so far as it is possible, the protective screens and shields.

The cold patch must also be thermally isolated from conductive heat leaks at the contact points where it is secured to the spacecraft. The instrument sensor which is being cooled by the radiant cooler will or course need to be coupled electrically to its electronic processor. These leads provide a virtually inescapable avenue for conductive heat leaks and there may also be some slight power dissipation in the sensor itself.

Figure 6.1 is a diagram of a representative, well-designed, radiant cooler capable of achieving 2 mW cooling at 80 K with a cold patch 45 cm^2, two states of cooling, a total cooler system, and a mass of 3.6 kg in a cylindrical space 112 cm in diameter and 55 cm long.

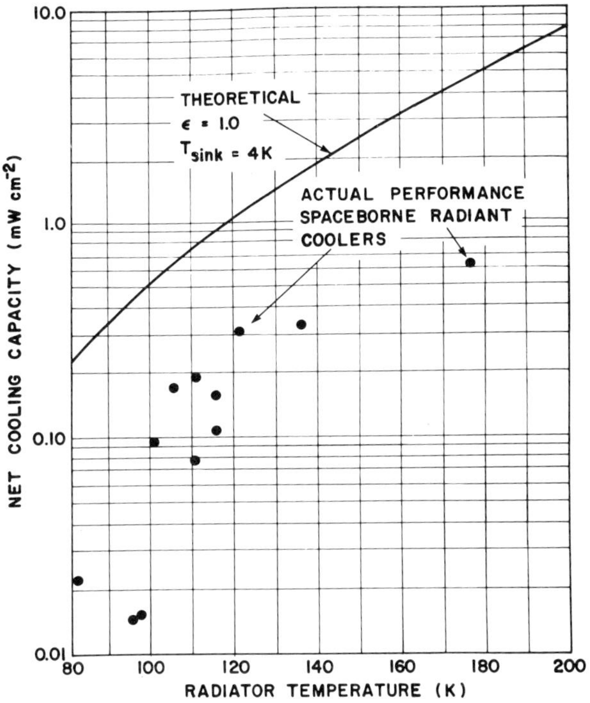

Fig. 6.2 Cooling capacity per unit area of cold patch as a function of cold patch temperature (after Donabedian 1978).

Figure 6.2 shows the net cooling capacity per unit area of the cold patch (mW/cm^2) as a function of the cold patch temperature in K (after Donabedian 1978, units converted). The upper curve shows the theoretical limit assuming an emittance $\epsilon = 1.0$ and a space sink temperature of 4 K The degree to which this theoretical limit was approached can be determined by reference to the marks below the upper curve indicating the performance of various radiant coolers in actual spacecraft service. It should be carefully noted that the vertical scale of Fig. 6.2 is logarithmic and that even the best of the practical coolers achieved only 30 per cent, approximately, of the theoretical limiting performance.

Doubling the temperature from 80 K to 160 K increases the theoretical cooling capacity from 0.25 to 4 mW/cm^2 approximately. This is the increase in capacity of 16 times to be anticipated from the fourth power relationship of the temperature in the fundamental equation given above. The upper limit of temperatures in the cryogenic range is generally accepted as 110 K but, of course, successful instrumentation applications do exist at higher temperature levels.

An excellent and extensive review of some contemporary development in aerospace radiative coolers was given by Kuyper (1992–1993).

6.3 Thermoelectric coolers

Consider two wires of different metals joined at the two ends. An electric storage battery connected in one of the wires causes a direct current to flow in the circuit. Due to thermoelectric effects one of the wire junctions will become cold, the other will become hot. This is the basis for operation of thermoelectric coolers.

Thermoelectric coolers have many attractive features. There are no moving parts, they operate without noise, the weight and volume are small, and the heating and cooling function can be easily reversed by reversing the polarity of the direct current.

If the battery is replaced by a galvanometer and the two junctions are maintained at a different temperature an electromotive force, the Seebeck voltage, will be developed and electric power can be developed.

Although potentially useful as a power system or cooling device this bimetallic, thermoelectric arrangement is best known as a *thermocouple* widely used for measuring temperature. One junction is maintained at a known temperature (melting ice) while the other end is located at the point whose temperature is to be measured. For a given temperature difference the Seebeck voltage is characteristic of the metal pair used and increases as the temperature difference between the junctions increases.

Thermoelectric phenomena were first investigated by the French chemist Jean Peltier as early as 1834. It was soon found that metal pairs were not suitable for practical use in thermoelectric power systems due to the high coefficient of thermal conductivity. It was not until the discovery of semiconductor materials following the Second World War that thermoelectric power systems and coolers became a practical reality.

Even with these developments the presently available materials limit the maximum temperature difference that can be achieved with a single-stage thermoelectric cooler to less than 160 °C. Thus from ambient temperatures of 300 K the minimum temperature achievable is about 140 K. At these lower temperatures the coefficient of performance is very low.

Figure 6.3 shows the typical form of a single-stage thermoelectric cooler utilizing p-type and n-type semiconductor materials with metal connecting straps at the hot and cold junctions. Figure 6.4 shows a

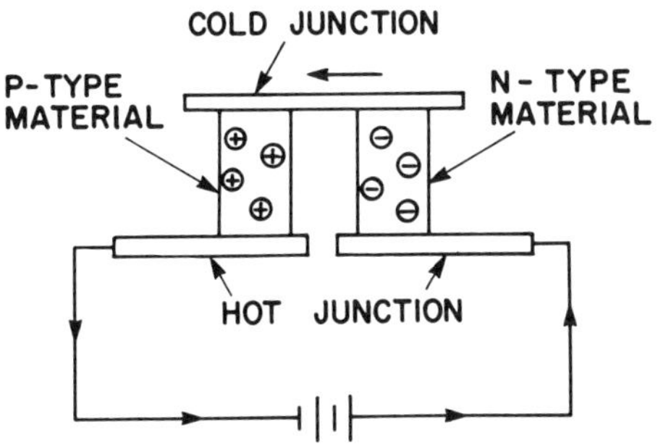

Fig. 6.3 Typical form of a single-stage thermoelectric cooler.

Fig. 6.4 Multistage thermoelectric cooler (courtesy Marlow Industries Inc., Dallas, Texas).

typical multistage thermoelectric cooler currently in production. This unit combines many identical thermoelectric couples connected in parallel to increase the refrigerating capacity. Several parallel stages are connected, one upon the other, in series (cascaded) to increase the temperature range beyond that achievable with a single-stage couple or to increase the coefficient of performance for a given temperature difference and heat pumping rate.

The most important performance parameter of a material for use in thermoelectric coolers is the figure of merit, Z. This is defined as

$$Z = \frac{S^2}{\rho k} \tag{6.2}$$

where Z is the figure of merit (K^{-1}), S is the Seebeck coefficient for a single material (volts per degree temperature), ρ is the electrical resistivity (ohm cm), and, k is the thermal conductivity ($W\ cm^{-1}\ K^{-1}$).

The highest values of the figure of merit are generally found in the semiconductor materials which combine a high Seebeck coefficient, a high electrical resistivity, and a moderate thermal conductivity. The average value of Z at room temperature is $2.5–3.0 \times 10^{-3}\ K^{-1}$ but this decreases markedly at low temperatures which dramatically decreases the coefficient of performance ratio at low temperatures.

The gross refrigerating capacity (or heat absorbed at the cold junction, sometimes called the Peltier effect) is:

$$Q_R = P_{ab} I \tag{6.3}$$

where

$$P_{ab} = S_{ab} T_R$$

and Q_R is the refrigerating capacity, P_{ab} is the Peltier coefficient between materials a and b ($W\ amps^{-1}$), I is the current (amps), S_{ab} is the Seebeck coefficient for the couple a, b ($V\ K^{-1}$), and, T_R is the temperature of the cold junction. The net refrigerating capacity is less than the gross refrigerating capacity by the losses due to resistance heating (I^2R), of which half is allocated to the cold junction and half to the hot junction, and the heat conduction loss

$$Q_c = K(T_H - T_R)$$

where R is the electrical resistance (ohms), K is the thermal conductance (WK^{-1}), and T_H is the temperature of the hot junction. If no external refrigeration load is applied then the minimum temperature achieved will be that value where the resistance heating and thermal conduction loss exactly equal the gross refrigerating capacity.

It can be shown that the maximum coefficient of performance (COP) of a single-stage thermoelectric cooler is:

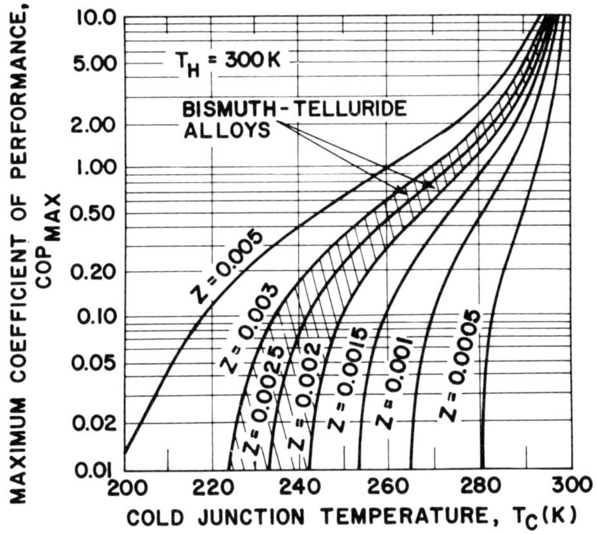

Fig. 6.5 Coefficient of performance of thermoelectric coolers as a function of cold junction temperature for different values of the figure of merit (after Donabedian 1978).

$$\mathrm{COP}_{\max} = \frac{\text{Heat lifted}}{\text{Power supplied}} = \left(\frac{T_{\mathrm{R}}}{T_{\mathrm{H}} - T_{\mathrm{R}}}\right) \frac{\sqrt{1 + 2T_{\mathrm{m}}} - (T_{\mathrm{H}}/T_{\mathrm{R}})}{\sqrt{1 + 2T_{\mathrm{m}} + 1}}$$

where T_{m} is the mean temperature $(T_{\mathrm{R}} + T_{\mathrm{H}})/2$.

Figure 6.5 shows the maximum COP as a function of the cold junction temperature T_{R} for different values of the figure of merit, Z. The shaded band between values of $Z = 0.002$ and $Z = 0.003$ are typical of the bismuth–telluride alloys used in commercial thermoelectric cryocoolers. Specific values of the figure of merit for typical p-type and n-type semiconductor materials are shown in Fig. 6.6 as a function of temperature.

Some improvement in the coefficient of performance may be won at the expense of complication through the use of multiple-stage units as shown in Fig. 6.7. It can be seen that the degree of improvement rarely justifies increasing the number of stages beyond four. These curves were calculated using the figure of merit for the materials given in Fig. 6.6 with an upper temperature of 300 K.

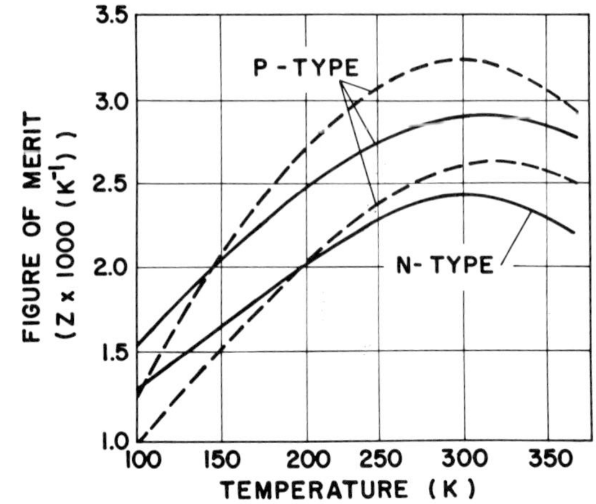

Fig. 6.6 Figure of merit for various p- and n-type semiconductor materials as a function of temperature (after Donabedian 1978).

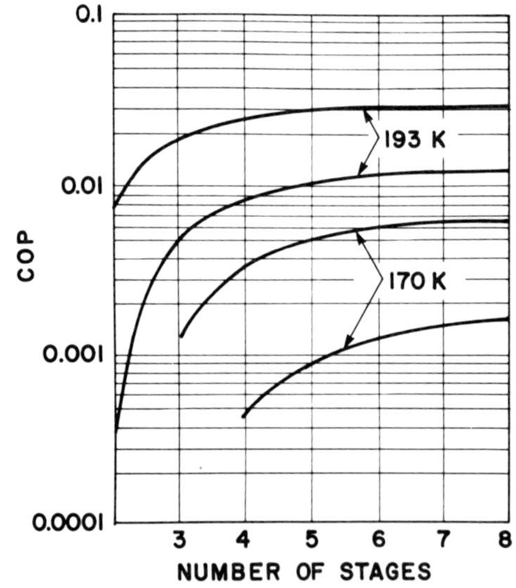

Fig. 6.7 Improvement in the coefficient of performance of thermoelectric coolers resulting from multistaging (after Donabedian 1978).

6.3.1 *Commercial applications of thermoelectric coolers*

Although not suitable in their present form for cryogenic cooling, thermoelectric coolers do have a wide and diverse field of application. One of the leading manufacturers in North America, Marlow Industries Inc., Dallas, Texas, list 54 different 'standard thermoelectric coolers' in their catalogue. Single-stage coolers achieve a maximum ΔT (temperature difference) of 67 °C with capacities ranging from 0.3 watt to 48 watts, two-stage machines can achieve ΔT of 96 °C, three-stage can achieve 110 °C, and six-stage units can achieve 133 °C. Applications include electro-optic and electronic components and a wide spectrum of volume cooling in space, military, domestic, commercial, and medical fields.

The advantages of thermoelectric cooling are enumerated in the Marlow catalogue as localized cooling, solid state reliability, temperature control above and below ambient, DC powered, no gas or refrigerant required, small lightweight package, operation in any orientation, high resistance to shock and vibration, operation in zero gravity, no moving parts, operation in high *g* levels, no acoustical noise, heating or cooling depending on current direction, no electrical noise, precision temperature control to less than 0.05 °C, and freedom from maintenance.

The above is a list of formidable advantages. Although thermoelectric coolers are by themselves unable to achieve cryogenic temperature with only ambient temperature cooling available their combination with other units should not be overlooked. They may, for example, provide a perfectly adequate first stage precooler for Joule–Thomson systems or for the other mechanical cryocoolers discussed above.

6.4 Sorption compressors

Whenever a gas comes into contact with a solid the intermolecular attractive forces cause some of the molecules to be retained or be *adsorbed* on the solid surface. The number of molecules adsorbed depends on the sorbent solid, the sorbate gas, the temperature and pressure at which the process occurs, and the surface area of the adsorbent solid. Certain solids, for example, charcoal and the aluminum silicate clay materials known as zeolites, have very porous structures. As a consequence they have very large surface area to weight ratios (over 1000 m^2/g) and so will adsorb large quantities of gas.

The amount of gas adsorbed increases as the temperature is reduced. A mass of sorbent solid that is saturated with gas at low temperature will desorb gas as the temperature is raised. If this process occurs in a closed container the pressure of gas being desorbed from the solid will

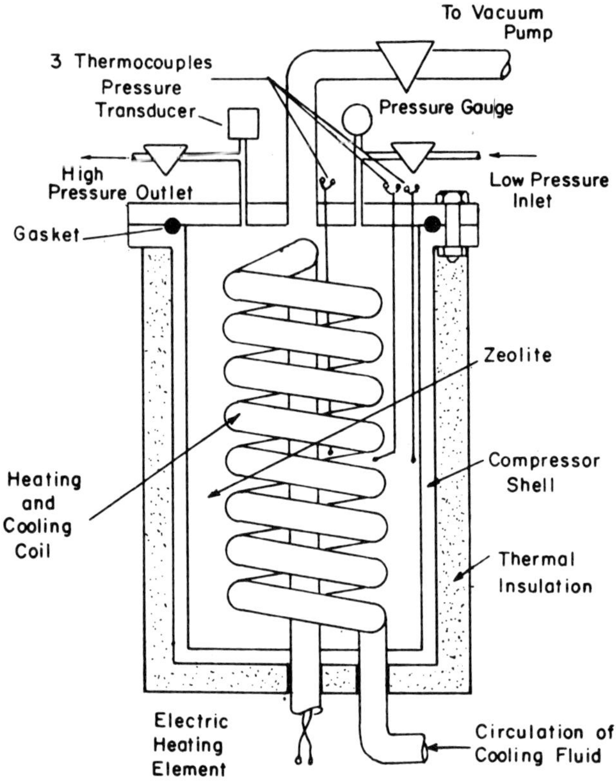

Fig. 6.8 Early experimental sorption compressor (after Hartwig 1978).

increase with the production of moderately high pressures (more than 100 atmospheres). If the compressed gas is released from the closed container and the sorbent solid is reduced in temperature it will then adsorb low-pressure gas admitted to the closed chamber. This process of thermal cycling can be continued indefinitely using an input of heat to accomplish the compression.

Figure 6.8 is a diagram of one early experimental sorption compressor described by Hartwig (1978). The closed and sealed chamber fitted with zeolite is equipped with (a) a dual-function heating/cooling element (electric resistance heating/water cooling), (b) inlet and outlet check valves, (c) the connections to a vacuum pump, and (d) instrumentation, pressure gauges, and thermocouple temperature measurement.

The thermal or adsorption compressor has no moving parts apart from automatic check valves to admit low-pressure gas to the chamber

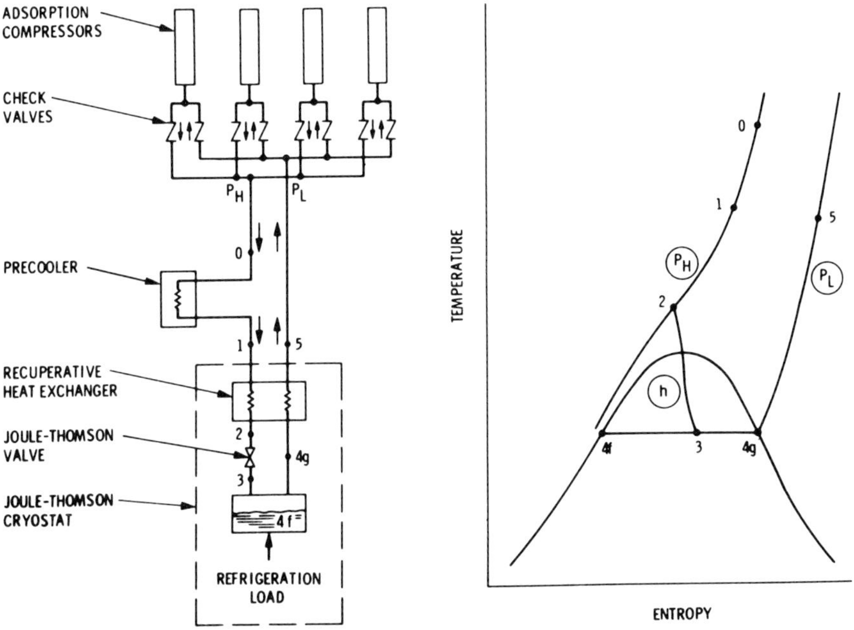

Fig. 6.9 Four-stage sorption compressor (after Bard 1978).

and release high-pressure gas from the chamber. The valves operate at a low frequency (once every two or three minutes) so can be expected to last many years. Sequential operation of a number of identical systems allows a substantially continuous flow of high-pressure gas to be delivered for expansion in a Joule–Thomson expansion nozzle or in the expansion engine of a Claude–Brayton or Gifford–McMahon cryocooler.

Figure 6.9 (after Bard 1987) shows a four-stage compressor generating a virtually continuous stream of compressed gas as part of a Linde–Hampson cryocooler with recuperative heat exchange followed by Joule–Thomson expansion of the compressed gas.

In a process that is similar to the adsorption process described above hydrogen is *absorbed* (chemically bonded) by intermetallic hydrides such as lanathium nickel ($LaNi_5$). The absorption of hydrogen is so intense, that, at room temperature, and a slight pressure, the density of the absorbed hydrogen is nearly twice the density of free liquid hydrogen at cryogenic temperatures.

An early exploration of desorption cooling at low temperatures (below 12 K) was reported by Daunt and Rosen (1972) and by Van Mal and Mijnheer (1972). Later Hartwig (1978, 1981) described the development of experimental sorption compressors at the University of Texas, at Austin. Hartwig's untimely death terminated this very promising program.

However, sufficient had been accomplished to arouse the interest of the aerospace community into the potential spacecraft application of sorption compressors for long-term, maintenance-free operation of vibration-frec, attitude and gravity insensitive compressors operating off a relatively low grade thermal input and without electromagnetic emissions.

The study and development of sorption compressors for these purposes was taken up at the Jet Propulsion Laboratory, Pasadena, California, for the National Aeronautics and Space Administration by the Aerojet Electronic Systems Division, Azusa, California, and by the Aerospace Corporation, El Segundo, California. Progress in the work was reported in a series of papers throughout the 1980s and early 1990s.

One paper by Jones (1985) summarized an extensive program of development of a $LaNi_5$ hydrogen absorption compressor capable of compressing hydrogen for subsequent liquefaction in a Joule–Thomson expansion. This paper is noteworthy for the extensive references to earlier work and also breadth of topics addressed including the production of solid hydrogen for 10 K refrigeration, the possible combination of the hydride sorption compressor with precooler refrigeration stages ranging from a Gifford–McMahon cryocooler, a passive radiant cooler, a thermoelectric cooler, and supplementary sorption compressors operating with charcoal sorbent and with methane and with nitrogen as the sorbate gas.

For this complex system, illustrated in Fig. 6.10, Jones proposed that solar heat could be collected at 425 K and used to power the four sorption compressors, all of which could, themselves, consist of three or four identical systems operating cyclically but out of phase.

Passive radiant coolers were proposed to cool:

(1) the charcoal sorbent of the methane and nitrogen compressors to 225 K;

(2) the metal hydride of the hydrogen compressors to 275 K.

The auxiliary thermoelectric cooler requires a relatively small input (30 W) of electric power to cool the pressurized methane, nitrogen, and hydrogen stream to 175 K. At this temperature the methane (at 26.5 atm) liquefies and, when expanded to 1.8 atm, cools to 120 K, cooling the nitrogen and hydrogen stages. The nitrogen at 24.8 atm pressure liquefies and on expansion to 1.4 atm, cools to 80 K thereby cooling the hydrogen stages. The hydrogen expands to a temperature of 18 K cooling the secondary hydrogen circuit. This expands in a final

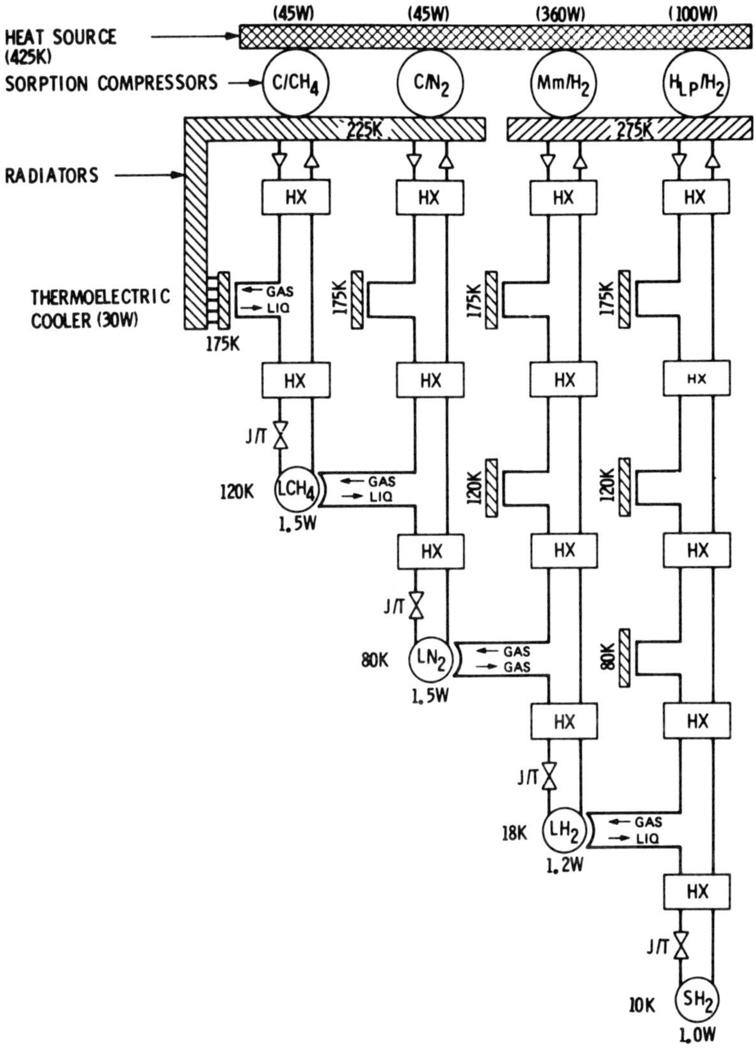

Fig. 6.10 Solar-heated compound 10 K hydrogen cryocooler (after Jones 1985).

Joule–Thomson nozzle to a pressure of 1.7 torr and a temperature of 10 K with the production of hydrogen 'snow'.

Jones estimates that a refrigeration capacity of 1 watt at 10 K would require a power input of 600 watts, most of it low-grade solar heating, making it the most efficient 10 K refrigeration system presently possible, primarily '... because it does not require the use of a very inefficient helium refrigeration cycle ...'. An enigmatic comment Jones does not enlarge upon.

Jones speculates on the possibility for a charcoal–helium sorption refrigeration stage to obtain temperatures of 4 K or so. However, his preliminary calculations indicated a power demand of several kilowatts per watt of refrigeration at 4 K. He favours the possibilities of an oil-free diaphragm compressor or a low-temperature (10 K) charcoal–helium sorption compressor but recognizes considerable development work is necessary to achieve this.

Jones asserts that 'Recent work at JPL has shown the possibility of achieving extremely long-life (10 years or greater) efficient and relatively simple refrigeration systems for obtaining 10 K temperatures and below.'

Another subsequent paper of exceptional interest from the Jet Propulsion Laboratory was given by Bard (1987). This reported on the design, development, and testing of a charcoal–nitrogen adsorption cryocooler operating between 100 and 120 K, pressure ranges of 9.3 to 57 atmospheres, and a specific power input of 629 to 310 watts per watt of refrigeration. Practical experience with the prototype provided the data for a proposed two-stage flight system design with:

(1) a refrigerator temperature of 80 K;

(2) pressure ratios of 1.35 to 20 and 20 to 100 atm;

(3) a specific power of 245 watts per watt;

(4) a system mass of 16 to 30 kg while incorporating a thermoelectric cooler having a specific power of 14.3 watts per watt, operating at 180 K;

(5) radiative coolers operating at 225 K; and

(6) a solar collector heat source operating at 450 K.

This could be viewed as the little brother of the much more complicated and ambitious 10 K cryocooler proposed by Jones and discussed above.

More recently Bard *et al.* (1993*a*,*b*) have discussed the development of a periodic 10 K sorption cryocooler and outlined the proposed 10 K sorption cryocooler flight experiment (BETSCE) scheduled for launch on the Space Shuttle in 1994. Contemporaneously Johnson and Jones (1993) have described the evolution of the 10 K periodic sorption refrigerator concept for long-term spacecraft use.

Activity has now spread to commercial developments with papers by Karperos (1987), Wade (1987), and Bowman *et al.* (1993) reporting on prototype hydrogen sorption refrigerator developments at Aerojet Electrosystems Company, Azusa, California.

Japanese experience in the field of absorption compressors was summarized by Matsubara *et al.* (1987) including the interesting possibility of including a high (sic—fast?) response hydride compressor in regenerative cryocoolers of the Gifford–McMahon, Stirling, and Vuilleumier genres. They found that a hydrogen sorption compressor with equilibrium pressure states of 12 atm at 80 °C and 4 atm at 40 °C could be cycled with a pressure response of about two seconds (as compared to the 16 times per second of a mechanical cryocooler running at 1000 revolutions per minute!). This pioneering paper opens the possibility of application of sorption compressors to the pulse tube and the development of cryogenic refrigerators having no moving parts at all.

6.5 Magnetic refrigerators

The following is a condensed extract from the paper by Kral and Barclay (1991). Dr Barclay, now a professor of engineering at the University of Victoria, British Columbia, Canada, has long been a researcher and advocate of magnetic refrigeration. His many contributions to the field will be found dispersed throughout the principal cryogenic literature over the past decade. The time for his definitive monograph on magnetic refrigeration is overdue.

6.5.1 *The magnetocaloric effect*

The entropy change, dS, experienced by a constant-volume magnetic material undergoing a change in temperature, T, or external magnetic field, H, is:

$$\mathrm{d}S = \left(\frac{C_{\mathrm{H}}}{T}\right)\mathrm{d}T + \mu_{\mathrm{o}}V\left(\frac{\delta M}{\delta T}\right)_H \mathrm{d}H \tag{6.4}$$

where C_{H} is the field-dependent heat capacity, μ_{o} is the permeability of free space, V is the volume, and M is the magnetization.

The temperature change experienced by a thermally isolated magnetic material when the material undergoes a change in magnetization induced by a change in external magnetic field is called the adiabatic temperature change. Under adiabatic conditions d$S = 0$, and equation (6.4) yields

$$\mathrm{d}T = -\left(\frac{T}{C_{\mathrm{H}}}\right)\mu_{\mathrm{o}}V\left(\frac{\delta M}{\delta T}\right)_H \mathrm{d}H\,. \tag{6.5}$$

Another quantity important to magnetic refrigeration following from Eq. (6.4) is the material's isothermal entropy change, the entropy change experienced at constant temperature in the presence of a changing field, given by

$$dS = \mu_0 V \left(\frac{\delta M}{\delta T}\right)_H dH . \qquad (6.6)$$

6.5.2 *Magnetic refrigeration cycle*

All magnetic refrigerators cool through the use of the entropy changes associated with magnetic field changes. The most common design cycle for magnetic refrigerators is the Carnot cycle. As larger operating temperature ranges and very large cooling powers are sought, regenerative cycles such as the Ericsson, Brayton, or active magnetic regenerator cycles are being used. These cycles are illustrated in Figs 6.11, 6.12, 6.13, and 6.14.

The Carnot cycle, shown in Fig. 6.11, is the simplest refrigeration cycle, but its use is limited by the adiabatic temperature changes that can be achieved by magnetic materials. The lattice heat capacity of the magnetic refrigerants used in these refrigerators increases rapidly with temperature. The adiabatic temperature change is limited by the increasing heat capacity as temperature rises, restricting Carnot cycle refrigerators to use at temperatures below about 20 K At temperatures above 20 K, and, for many applications below 20 K, regenerative cycles are used.

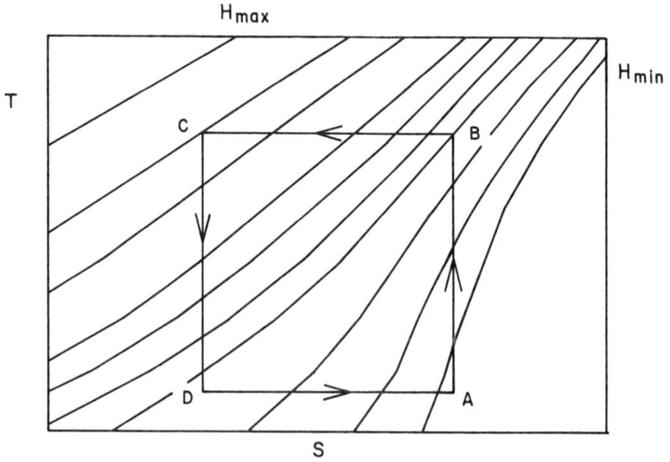

Fig. 6.11 Magnetic Carnot cycle with magnetic material isofield lines (after Kral and Barclay 1991).

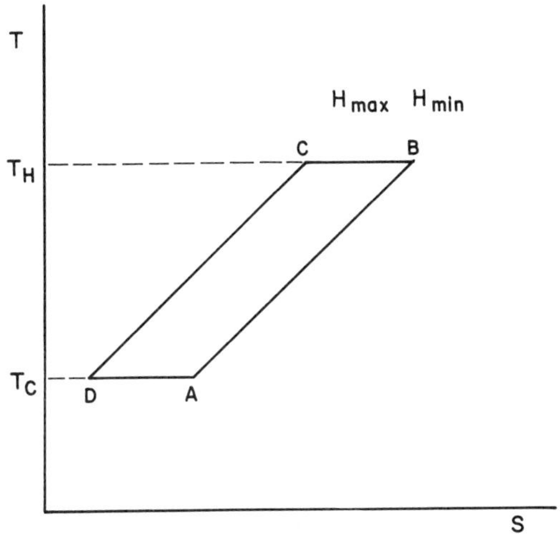

Fig. 6.12 Magnetic Ericsson cycle: isothermal heat transfer and isofield regeneration (after Kral and Barclay 1991).

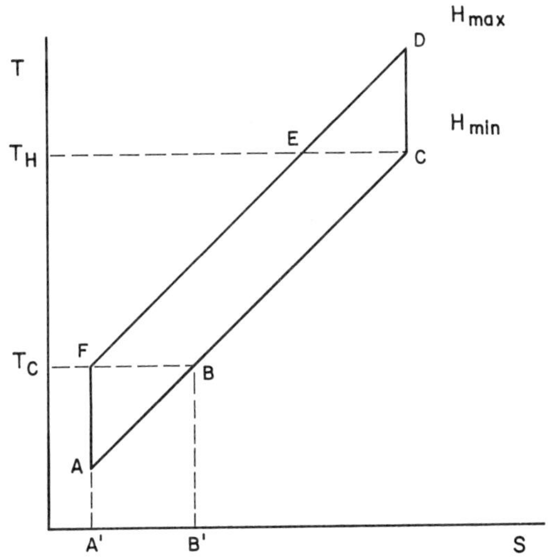

Fig. 6.13 Magnetic Brayton cycle: isofield heat transfer with isofield regeneration (after Kral and Barclay 1991).

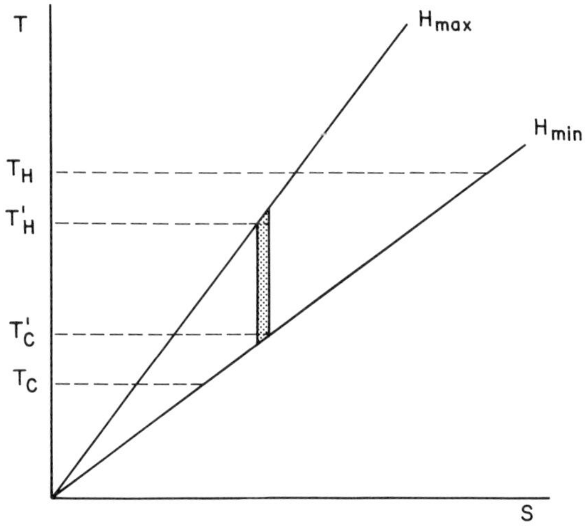

Fig. 6.14 AMR cycle: elemental Brayton cycle operation between T'_H and T'_C is shown (after Kral and Barclay 1991).

Regenerative cycles allow much larger temperature spans and potentially much higher cooling powers than Carnot cycles. Figure 6.12 presents a regenerative magnetic Ericsson cycle. In this cycle, for high efficiency, the heat transferred from the regenerator to the magnetic material as it warms on path AB must be exactly balanced by the heat transferred to the regenerator as the magnetic material cools along path CD. Because the heat transferred during regeneration is simply the area between the entropy axis and the path followed during regeneration, CD or AB, for an ideal Ericsson system, the magnetic material constant-field curves in the T–S plane must be parallel straight lines.

In both the magnetic Carnot and magnetic Ericsson cycles, heat transfer occurs isothermally while the material is being magnetized or demagnetized. For an ideal cycle, without irreversibilities that distort the cycle, the cooling power is given by

$$Q_{\text{ideal}} = T_c \, \Delta s_T f m \tag{6.7}$$

where T_c is the cold source temperature, Δs_T is the isothermal entropy change per unit mass for the demagnetization occurring at the cold sink temperature, f is the cycle frequency, and m is the mass of the magnetic material.

The magnetic Brayton cycle is another possible regenerative cycle. In the ideal magnetic Brayton cycle, heat is absorbed at a constant magnetic field. The cooling power can be calculated easily for the ideal case, shown in Fig. 6.13, because the heat absorbed during each cycle is the area of the trapezoid AA′BB′. If ΔT_s is the adiabatic temperature change experienced by the material during the demagnetization, and Δs_T is the isothermal entropy change (per unit mass) that would occur if the material underwent an isothermal demagnetization between points E and B, the cooling power is seen to be

$$Q_{\text{ideal}} = T_c \Delta s_T fm - \left(\frac{1}{2}\right) \Delta s_T \Delta T_s fm \,. \tag{6.8}$$

The cooling power for an ideal Brayton cycle is seen to be reduced from that of an ideal Carnot or Ericsson cycle operating under similar conditions.

Another important cycle in magnetic refrigeration is the active magnetic regenerator (AMR) refrigerator cycle. For this cycle, in the T–S plane (Fig. 6.14), the magnetic material ideal isofield lines are lines that are not parallel, but rather straight lines that intersect at $T = 0$ K. This cycle also requires a second medium to effect heat transfer. In an ideal AMR, each element of the magnetic material is carried through a small ideal Brayton cycle which spans only part of the total temperature range spanned by the refrigerator. Over a cycle, upon magnetization, all of the elements are raised to their highest temperature. The second medium, typically fluid helium, is warmed from the cold source temperature to slightly above the hot sink temperature by heat absorption from the magnetic material. The second medium then rejects heat to the hot sink. All the elements are then cooled to their lowest temperature by demagnetization. The second medium is cooled to slightly below the cold source temperature as it rejects heat to the magnetic material elements. Heat is then absorbed from the cold source by the second medium and the cycle begins again. In this way, the heat rejected by the second medium as it cools is absorbed by the elemental Brayton cycle refrigerators when they are at the low-temperature part of their cycles. Heat is rejected to the second medium when the elemental Brayton cycle refrigerators are at the high-temperature part of their cycles. In a superficial way, this process resembles internal regeneration; the process is more precisely understood as the action of an ensemble of elemental Brayton cycle refrigerators acting in parallel to warm and cool a second medium to effect heat transfer from the cold source to the hot sink. The cooling power of an AMR depends directly on the capacity of the second medium to absorb heat from the cold source. If m is the mass of the second medium, c_p is its specific heat, f is the cycle frequency, and ΔT is the temperature difference between the cold source and the second medium, the cooling power is given by

$$Q = mc_p f \Delta T . \tag{6.9}$$

Because the second medium is cooled by Brayton cycle refrigerators, the AMR is intrinsically limited to efficiencies less than those of a Carnot cycle.

The description of the AMR refrigeration cycle given above is correct under conditions where the thermal mass of the magnetic material is much greater than that of the second medium. When the thermal mass of the second medium is greater than that of the magnetic material, the ideal isofield lines in the T–S plane are parallel, to ensure balanced regeneration.

In their review Kral and Barclay (1991) include a table of data for 30 different magnetic refrigerators and conclude:

'The cooling powers of the magnetic refrigerators now being designed, fabricated, and tested are extremely large, some with cooling power in the tens of kilowatt range. These designs are all extremely compact. In one design unit a cooling power of approximately 10 kW at 20 K has the refrigerant confined to an annular cylinder less than 2 m in diameter and less than 0.5 m in height. Modeling predictions for this design also indicate a higher efficiency than competing gas systems. This result, while remarkable, is not unique.

Magnetic refrigeration is an emerging technology in a market place long dominated by gas-cycle technology. As the properties of magnetic refrigerators become better known through actual validated models and performance, other incorrect preconceptions, similar to the small cooling power assumption, will be identified and clarified. Recent successes in both conductive and convective designs and new, ambitious designs continue to reveal the interesting properties of this technology.

References

Bard, S. (1981). Development of an 80–120 K Charcoal–Nitrogen Adsorption Cryocooler. *Proceedings of the Fourth International Cryocooler Conference* (ed. G. Green, G. Patton, and M. Knox), David Taylor Naval Ship R. and D. Center, Annapolis, Md.

Bard, S., Fujita, T., Wade, L., Rodriguez, J., and Wu, J. J. (1993(a)). Development of a Periodic 10 K Sorption Cryocooler. *Proceedings of the*

Seventh International Cryocooler Conference (ed. J. Ludwigsen), Kirtland AFB, Santa Fe, N.M.

Bard, S., Cowgill, P., Wu, J. J., Rodriguez, J., Wade, L., Gehrlein, M., and Von Der Ohe, W. (1993). 10 K Sorption Cryocooler Flight Experiment (BETSCE). *Proceedings of the Seventh International Cryocooler Conference* (ed. J. Ludwigsen), Kirtland AFB, Santa Fe, N.M.

Bowman, R. C. Jr., Freeman, B. D., Labor, D., Lynch, F. E., Marmaro, R. W., and Wade, L. A. (1993). Evaluation of a Prototype Hydride Compressor for Periodic Hydrogen Sorption Cryocoolers. *Proceedings of the Seventh International Cryocooler Conference* (ed. J. Ludwigsen), Kirtland AFB, Santa Fe, N.M.

Daunt, J. C. and Rosen, C. Z. (1972). Desorption Cooling Below 12 K Using He^4 Desorbed From Synthetic Zeolite. *Cryogenics*, June, Butterworths Scientific Publishers, Guildford, UK.

Donabedian, M. (1978). Cooling Systems. Chapter 15 in *The Infrared Handbook* (ed. L. W. Wolfe and G. J. Zissis), Infrared Information and Analysis Center, Environmental Research Institute of Michigan, Ann Arbor, Mich.

Hartwig, W. H. (1978). Cryogenic Applications of Closed Cycle Mechanical and Adsorption Refrigeration. In *Applications of Closed Cycle Cryocoolers to Small Superconducting Devices* (ed. J. Zimmerman and T. Flynn), NBS Special Publication 508, US Government Printing Office.

Hartwig, W. H. (1981). Requirements For and Status of 4.2 K Adsorption Refrigerator Using Zeolites. In *Refrigeration for Cryosensors and Electronic Systems* (ed. J. Zimmerman, D. Sullivan, and S. McCarthy), NBS Special Publication 607, US Government Printing Office.

Johnson, A. L. and Jones, J. A. (1993). Evolution of the 10 K Periodic Sorption Refrigerator Concept. *Proceedings of the Seventh International Cryocooler Conference* (ed. J. Ludwigsen), Kirtland AFB, Santa Fe, N.M.

Jones, J. (1985). Hydride Absorption Refrigerator System for 10 K and Below. *Proceedings of the Third Cryocooler Conference* (ed. R. Radebaugh, B. Louie, and S. McCarthy), NBS Special Publication 698, US Government Printing Office.

Karperos, K. (1987). Operating Characteristics of a Hydrogen Sorption Refrigerator, Part 1: Experiment Design and Results. *Proceedings of the Fourth International Cryocooler Conference* (ed. G. Green, G. Patton, and M. Knox), David Taylor Naval Ship R. and D. Center, Annapolis, Md.

Kral, S. F. and Barclay, J. A. (1991). Magnetic Refrigeration: A Large Cooling Power Cryogenic Refrigeration Technology. In *Applications of Cryogenic Technology*, Vol. 10 (ed. J. P. Kelley), Plenum, New York.

Kuyper, D. (1992). Radiative Cooler for the Pressure Modulator IR Radiometer. *Cryogenic Optical Systems and Instruments*, V. SPIE, **1765** (23), July.

Kuyper, D. (1993). A Radiative Cooling System for the EOS Stratospheric Wind Infrared Limb Sounder. *Proceedings of the Seventh International Cryocooler Conference* (ed. J. Ludwigsen), Kirtland AFB, Santa Fe, N.M.

Matsubara, Y., Kaneko, M., Suzuki, J., and Hirosawa, K. (1987). High Response Hydride Compressor for Regenerative Cryocooler. *Proceedings of the Fourth International Cryocooler Conference* (ed. G. Green, G. Patton, and M. Knox), David Taylor Naval Ship R. and D. Center, Annapolis, Md.

Van Mal, H. and Mijnheer, A. (1972). Hydrogen Refrigerator for the 20 K

Region with a $LaNi_5$ Hydride Thermal Absorption Compressor for Hydrogen. *Proceedings of the Fourth International Cryogenic Engineering Conference*, IPC Science and Technology Press, Guildford, UK.

Wade, L. A. (1987). Operating Characteristics of a Hydrogen Sorption Refrigerator, Part 2: A Comparison Between a Second Order Analysis and Empirical Data. *Proceedings of the Fourth International Cryocooler Conference* (ed. G. Green, G. Patton, and M. Knox), David Taylor Naval Ship R. and D. Center, Annapolis, Md.

7

MECHANICAL CRYOREFRIGERATORS FOR SPACECRAFT

Dr Otto C. Ledford Jr,

This chapter was contributed by Dr Otto C. Ledford, Jr. It is an updated revision of the paper of the same title presented by Dr Ledford at the Cryogenic Engineering Conference, 1987, but which was not published in the proceedings of that conference, that is, Advances in Cryogenic Engineering, *Vol. 33, (ed. R. Fast) Plenum.*

Abstract

There are many missions performed in space which require cryogenic temperatures for their operation. Although many of these requirements can be met by stored cryogens or radiators, mechanical cryorefrigerators will often allow a lighter, more flexible system. In many cases, the combination of high loads, low temperatures, and long mission duration or continuous operation makes the availability of a long-life cryorefrigerator a requirement for the performance of the mission. This paper gives a historical perspective of the development of mechanical refrigerators for spacecraft, describes some current programs, and presents some opinions regarding the present and near-term state of the art.

7.1 Introduction

One of the primary applications of space has been as a location for sensors to take advantage of the viewing geometry or absence of atmospheric disturbances. Missions which have been performed or proposed include remote viewing of the Earth and its atmosphere, astronomy, and the detection and tracking of objects of military interest. Many of the instruments required to perform these missions operate satisfactorily only when cooled to cryogenic temperatures. Examples of this are detectors designed to respond to gamma radiation or light waves in the infrared portion of the spectrum. In the case of infrared detectors, there is an inverse relationship between the maximum wavelength and the temperature of operation. Hence, whereas a sensor on a weather satellite responding to relatively short-wave infrared radiation will operate satisfactorily in the region of 200 K, a sensor for an astronomy satellite responding to very long wavelength sources from the cold space background must be cooled to a temperature near absolute zero. For example, the IRAS (Infra-Red Astronomy Satellite) focal plane, which had a response extending to 120 microns, was cooled to 2.7 K by superfluid helium. Often, the optics of these systems must also be cooled to cryogenic temperatures to reduce noise, with the baffles around the sensor cooled to somewhat warmer (liquid nitrogen or above) temperatures to reduce the loads at the colder temperatures. There have recently been other applications requiring cryogenic temperatures proposed for space-based systems. These include superconducting electronic devices and the long-term storage of large amounts of cryogenic fluids.

At relatively high temperatures and low loads, as with the weather satellite mentioned above, the cooling can be provided by a radiator oriented towards space. Stored cryogens in either the liquid or solid state have traditionally been used to provide cooling at lower temperatures. However, as the heat load and/or the desired mission duration increases, the weight of such a stored cryogen cooling system will also increase. It has long been recognized that if a reliable, efficient refrigeration system capable of reaching cryogenic temperatures was available, it would be possible to realize substantial system weight savings for many missions. Accordingly, there have been efforts to develop such devices since the earliest days of the space program. The development of cryogenic refrigeration technology has been recently accelerated by the desire to perform missions of such heat loads and duration that passive or stored cryogen cooling are not practical.

7.2 Flight experience

The first known use of a mechanical refrigerator in space was an experiment known as RM-19 launched by the US Air Force in 1970. This incorporated two single-stage Stirling refrigerators built by the Malaker Corporation operating in parallel. This machine had a bevel gear drive and a crank mechanism reciprocating the compression and expansion pistons, which used rubbing seals. It was a modification of a ground-based tactical cooler. The cooling load was 1.7 watts at 100 K. The refrigerator operated satisfactorily for the six-month mission, although it showed the steady performance degradation characteristic of wearing machines. There was a sharp decrease in performance early in the mission, but this is believed to be due to contamination of some low-emissivity spacecraft surfaces (thus increasing the heat load) rather than a problem with the cooler itself. Variants of this design producing 1 watt at 80 K also flew on the Skylab in 1973 and several Shuttle missions.

In 1971, a two-stage VM cooler built by Hughes flew as part of an Air Force celestial IR mapping experiment on the STP 71–2 spacecraft. The cooler, a dry-lubricated design, produced 0.15 watts at 15 K and 3.5 watts at 55 K. The cooler performed very erratically the first few days in orbit. The power consumption was much higher than predicted and the required temperatures could not be maintained. It was never clear whether this was due to a fault in the refrigerator or other effects. One theory was that water vapour in the sensor caused the problems until it outgassed. The cooler did settle down to satisfactory performance for about two weeks until the heat rejection system failed, probably due to a leak in a rotary joint that caused loss of coolant.

In 1979 another STP spacecraft, P78–1, was launched containing a mechanical refrigerator as part of its payload. Four rhombic drive grease-lubricated Stirling refrigerators cooled the sensors of two gamma ray spectrometers. Philips Laboratories began developing these coolers under DARPA sponsorship in 1973. Each refrigerator was designed to provide temperatures of 140 K on the first stage and 77 K on the second stage with heat loads of 1.5 watts and 0.3 watt respectively. They weighed 7.1 kg each, and required 30 watts input power. The original design called for the working volume to be separated from the lubricated mechanisms by a metal bellows having theoretically infinite life. However, bellows failed regularly during development tests and were omitted from the flight coolers. One of the coolers began losing helium at a significant rate immediately after launch. This was thought to be due to a leak in the crankcase or the refill valve not reseating properly. The other three coolers lost helium at a lower, but steady rate. This was attributed to diffusion through Viton seals and the magnesium casing. Although the thermal performance degraded as a result of the pressure loss, the coolers functioned perfectly mechanically. They were turned off and on many times and allowed the experimenters a great deal of operational flexibility. Instrument performance was maintained after cooler performance had degraded by operating the primary and redundant cooler together. The refrigerators were still operating and the payload was still taking data when the spacecraft was used as the target for a successful test of an Anti-Satellite interceptor in 1985.

The Russians have also flown a number of cryogenic refrigerators on their Salyut space station. Details of these devices are rather sparse. However, there is reason to believe that it may have been a hybrid configuration, with two Stirling stages precooling a Joule–Thomson (JT) cycle.

The next application of a mechanical cryocooler in space is scheduled to be an experiment on which a pair of split-Stirling coolers weighing 2 kg and drawing 30 watts each are to be mounted in phase opposition so as to minimize vibration. These are space qualified models of a production tactical cooler producing 1 watt at 80 K and guaranteed for 2500 hour operation by Magnavox. They were originally developed by Philips in the Netherlands. They have rubbing seals, and performance thus degrades as wear occurs. However, life tests performed by Aerojet indicate there will be adequate performance at the end of the one-year experiment life, since the system has excess initial capacity. This is an example of how a small, light refrigerator of less than infinite life can be very satisfactory in an experimental application.

The next generation of machines that are suitable for such application may be found in the family of split-Stirling coolers being developed by a number of sources in the United Kingdom, following the work of Davey at Oxford. These use flexure supports and clearance seals to avoid the rubbing wear that limits the life of similar small machines. These machines are being carried as the baseline design for a number of proposed space missions. The original machine, designed for a payload on the NASA-Goddard UARS satellite provides 0.75 watts at 80 K. The concept has been extended to multiple stages and higher loads in development units.

7.3 Vuilleumier refrigerators

In the early 1970s, there was a great deal of interest in the development of Vuilleumier (VM) refrigerators for space applications. This cycle, which is fundamentally a Stirling refrigerator having a mechanical compressor engine, was attractive because it could directly use thermal energy to produce refrigeration. At the time there was much interest in using radio-isotope heat sources in spacecraft. However, the political climate in recent years has dictated that electric heaters drawing power from solar cells be used instead. This is inherently less efficient than using the electrical power to drive a compressor motor. There must be advances in the technology of concentrating and transporting solar thermal energy or a modification of policy regarding radio-isotopes before the advantages offered by this concept can be fully realized.

Several contractors were funded to various levels of development during this period. For example, Garrett Airesearch built two dry-lubricated VM machines producing 5 watts at 65 K under NASA/GSFC funding and Philips Laboratories built a three-stage oil-lubricated rhombic drive VM under US Air Force funding. However, it is Hughes Aircraft

Company that has shown the greatest commitment to the VM cycle, and which is usually associated with it. Hughes built three copies of the so-called High Capacity (Hi-Cap) VM cooler for the Air Force beginning in 1970. This was a dry-lubricated design selected in preference to the Philips oil-lubricated concept. Three more copies of a space qualified derivative were built for the Space Infra-Red Experiment (SIRE) program. These six Hi-Cap and SIRE coolers provided the hardware for a continuing test and development program which culminated in one of the SIRE machines being run for two-and-one-half years at double speed, arguably the equivalent of 5 years normal operation. The machine did leak badly due to the many bolted connections, so a helium makeup system or new packaging would be required to reach such a life in space service. This test, which represents the present state of the art of demonstrated lifetime, is all the more remarkable when it is considered that the original lifetime requirement was six months.

This duration of operation was achieved only as a result of a dedicated product improvement effort, which can be appreciated by examining the mechanical details of this design. The linear motions of the hot and cold end displacers are coupled by a rotary linkage with dry-lubricated ball bearings which are subject to wear, and flex pivot connectors which were subject to fatigue failure. Both the hot and cold ends had rubbing fluorocarbon seals to contain the pressure differential and rubbing riders to take out the side forces resulting from the rotary to linear motion. All of these rubbing components showed a very poor initial wear life history and required the development of new materials and geometries. There were also extensive problems with gaseous contamination of the helium working fluid. The most recent of these was during an accelerated wear test run in parallel with the life test mentioned above. The debris generated by seal and rider wear did not interfere with the moving mechanisms as had been feared. However, it broke down in contact with the hot end surfaces to form fluorine compounds which reacted with other organics to form carbon monoxide, which froze out in the cold end. The machine had to be purged and refilled with clean helium several times in order to complete the test. There is some indication that this phenomena was also occurring to a lesser degree on the two-and-a-half-year life test.

7.4 The quest for long life

As plans for space-based missions grew more ambitious and the cost of the systems supported grew more expensive, the lifetime and reliability requirements for cryogenic refrigerators were increased accordingly. It was recognized that the wearing and fatiguing components of the designs discussed above would ultimately limit the life of mechanical refrigerators to less than the steadily growing requirements, despite the best product improvement efforts. More critically, the lifetime of wearing components cannot be theoretically assessed, but can be characterized only by adequate amounts of experimental data. This means that a reliability assessment satisfying potential users can be obtained only by running many machines to failure, a procedure which would have to be repeated for each new configuration. Since a typical minimum system reliability requirement for a cryocooler is currently 0.95 for a five-year life, the amount of failure data necessary to statistically characterize a wearing machine is quite substantial. Neither the time nor the money would be available to meet this requirement.

The solution to this problem has been known for some time. Wearing failure must be eliminated by avoiding rubbing mechanisms through the use of non-contacting seals and bearings. Fatigue failure must be eliminated by keeping stresses low. Contamination must be eliminated by using only non-organic materials within the refrigerator and using extreme care in purging and filling the working fluid. Leakage must be reduced by hermetically sealing the working volume. A philosophy of designing out the mechanical failure modes can result in a system in which the life and reliability is dominated by the electronics. Standard, accepted methodologies exist for predicting electronic component reliabilities, largely based upon piece part count and environment. With the mechanical failures, which cannot be well characterized theoretically, contributing a relatively small part of the overall failure analysis, a reasonable estimate of the lifetime and reliability of the refrigerator can be made without the need for the extensive test program. A test program sufficient to discover and correct design errors and verify that the non-wearing components are in fact non-wearing can thus replace the need to run large numbers of machines to failure. It is the implementation of this philosophy that has proven difficult.

Three basic non-contacting bearing systems have been used to prevent wear in the moving mechanisms of cryogenic refrigerators: hydrodynamic gas bearings, hydrostatic gas bearings, and magnetic suspension. Typically, clearance seals are used with reciprocating components whilst labyrinth seals are utilized with rotating components. A clearance seal relies upon a relatively narrow gap, with respect to its length, to maintain pressure differentials. Labyrinth seals offer a tortuous path to restrict flow. Both concepts have greater leakage than rubbing seals, but this penalty to system performance is accepted in order to obtain the required life and reliability.

The USAF set out to obtain a long-life spacecraft cryogenic refrigerator in 1962. A request for proposal (RFP) for a non-reciprocating closed-cycle cryocooler led to a reversed Brayton turbo-cooler which produced 2 watts at 80 K being built by Garrett Airesearch. This machine used conical foil hydrodynamic bearings to avoid wear. In 1970, a competition for a similar machine producing 1.5 watts at 12 K and 40 watts at 60 K was won by General Electric. Their reversed Brayton cycle turbo-machinery-based concept used tilting pad gas bearings. This effort was terminated in 1974 due to the high input power required for the small loads then under consideration.

In response to a parallel RFP for a reciprocating closed-cycle cryocooler in 1962, A. D. Little was given a contract for a concept that was also based upon the reversed Brayton cycle and hydrodynamic bearings, but in a quite different implementation. It was called a rotary reciprocating refrigerator, or Rcube for short; so named for its method of operation. A piston is rotated within a cylinder to establish a hydrodynamic bearing with the working fluid, and then reciprocated to obtain the compression and expansion necessary to carry out the thermodynamic cycle. Integral ports which are covered and uncovered by the phasing of the rotation and reciprocation control the gas flow. Although all of its implementations to date have been reverse Brayton cycles, it can in principle be used with any thermodynamic cycle. In fact, A. D. Little performed a design study of a three-stage Stirling Rcube in 1979. A. D. Little was also chosen to develop a two-stage 12 K/60 K refrigerator in 1970. This hardware achieved full system status after several funding interruptions and some technical difficulties.

The technical challenge of this concept derives from the fact that gas viscosity is a direct function of absolute temperature. At the design point near 10 K, the viscosity of the helium is so low that a radial clearance of 50 millionths of an inch between the piston and cylinder was required to maintain adequate bearing stiffness. To achieve this tolerance along the entire length of the piston and cylinder was a challenge in itself. This tolerance then had to be maintained as the assembly cooled down from ambient to the operating temperature.

The expansion engines were originally fabricated from Ferrotic-C, a matrix of tungsten carbide in tool steel. This had been selected because of its history as a material for gas gyros. However, in this application it was found to function as a bi-metallic switch. As the piston cooled down, it bowed slightly until it contacted the cylinder wall during rotation, well above the design point temperature. This problem was discovered in 1982, and it took another four years to find a replacement material and fabricate a new cold expansion engine. The first-stage expansion engine and the compressor operated satisfactorily. Since they operated at higher temperatures with correspondingly higher viscosities, they had much larger bearing clearances, and were physically much thicker, stiffer, and resistant to distortion.

The engine material had requirements which were conflicting. It had to be homogenous and thermally stable, which implies a simple alloy, and it had to be very hard so that it could run on itself without damage during those moments at start-up and shut-down where the gas bearing was not established. A complex alloy is usually required to achieve this hardness. Coatings and platings were avoided because of the fear of small chips coming loose at edges which could cause catastrophic damage in the small clearance. An extensive materials evaluation and test program did eventually identify a suitable material, and a new engine was fabricated.

This engine was tested as part of a complete closed-loop refrigerator, along with the warm-stage engine, compressor, and control electronics that had been fabricated earlier. This system produced 2.4 watts at 11.5 K and 10 watts at 59 K under full load. Temperatures of 9.4 K and 53.9 K were reached under no-load conditions. Only one of the two compressor modules was used during the no-load test, and there is every reason to believe that even lower temperatures would have been achieved had the full design mass flow been available. The complete system was run for 3187 hours, and post-test examinations revealed no sign of wear.

NASA was also moving towards the goal of long-life cryogenic refrigeration in the 1970*s*. They funded two competing Stirling cycle refrigerators for a requirement of 5 watts at 65 K. Both concepts utilized linear motor drives to achieve purely linear motion and reduce radial loads. Philips Laboratories used the active magnetic bearing suspension technology that had been pioneered at NASA/GSFC and ERG Inc. used gas bearings. The Philips concept was slightly more advanced at the end of phase I, when budgetary limits prevented the continuation of dual efforts. An engineering model of the Philips design was built and met all performance specifications. It operated in the Philips Laboratory in Briarcliffe Manor, N. Y. By 1989, it had logged over 43,000 hours (approximately 5 years) of refrigerator operation and over 60,000 hours of levitation on its magnetic bearings without mechanical failure or degradation of performance. Since laboratory-type control electronics are used, electronic components do have to be replaced periodically.

A flight qualified version of this refrigerator was being built by Philips in 1989. This machine was projected to weigh 70 kg, plus electronics, and require 185 watts of input power. Philips Labs, Briarcliffe Manor is no longer in the cryocooler business.

In 1981, DARPA initiated a program to extend this technology to lower temperatures. It funded the design and development of a two-stage linear magnetic suspension Stirling refrigerator designed to meet the requirements of the Mini-HALO sensor. A prime motivation for this was that such a cooler would require about a third the input power of the baseline reverse Brayton turbo-refrigerator. Magnavox Government Electronics Group, the manufacturing and commercial adjunct of Philips, was given a contract to build and test two refrigerators. Beginning in 1982, the contract was afflicted by increases in burden rates. There was also cost growth resulting from funding interruptions and design changes resulting from the Preliminary Design Review. In 1983, the program was transferred from DARPA to the USAF as part of the general restructuring of technology programs resulting from the Strategic Defense Initiative. This transition was quite traumatic, as the two agencies had quite different contractual standards, reporting requirements, drawing and quality control standards, and so on. This change in requirements led to more cost growth. This cost growth led to a series of program downscopes. The program was first reduced from two machines to one. There was then enough money to build the machine but not test it. Finally, the dynamic balancer was deleted. In 1985, as the final fabrication and assembly was taking place, it was discovered that certain operations which had been scheduled to take place in parallel could only be done in series due to a unique vendor capability. This resulted in a schedule slip with an associated cost increase. The government chose not to fund this incremental cost, and the program was thus terminated just short of completion. The program was relatively trouble-free technically, as all major technical issues had been solved and demonstrated on earlier programs.

In the mid-1970s, DARPA was interested in the concept of a very large surveillance system which could stare at the entire earth from geosynchronous orbit: the HALO (High Altitude Large Optics). Competitive studies showed that a turbo-machinery refrigerator was best suited for providing the large cryogenic cooling requirement of this system. A flight demonstration of subscale elements of the technology was planned, called Mini-HALO. It had optics somewhat greater than 1 metre in diameter and a large mosaic focal plane of mercury–cadmium–telluride with long-wave infrared (LWIR) response. Garrett Airesearch was selected to build the reverse Brayton cycle turbo-refrigerator to provide the cryogenic cooling of the optics and focal plane. As it turned out, the flight program was cancelled in 1983 when it was found that the cryogenic refrigerator was the only one of the payload technology elements that was able to meet its design and development requirements.

The Mini-HALO refrigerator uses a two-stage turbo-compressor rotating at 160,000 r.p.m. and two stages of expansion on a single shaft rotating at 200,000 r.p.m. The expansion turbines drove an alternator, and the extracted energy was fed back into the main power bus. Contact and wear of the moving parts was avoided by use of hydrodynamic foil journal bearings and a hydrostatic thrust bearing. As a consequence of the small size and high frequency of the moving parts, the vibration caused by this unit's operation is almost immeasurable. This unit was designed and tested to launch environment standards. It weighs 100 kg and requires 2600 watts input power. It suffered the usual technical and programmatic difficulties, but met its design requirements in the end. It is currently running a life demonstration test in Torrance, California. This test, which began in May 1989, is scheduled to last for 20,000 hours. This machine has logged a total of 9291 hours of operation prior to this test. The longest uninterrupted run was 7401 hours on a previous life demonstration test which concluded in June 1987.

7.5 The AMSCR program

In 1984, there was a high perceived risk in the two cryogenic refrigerator programs being supported by the Air Force. It was recognized that the wearing nature of the Hughes VM unfavorably affected its life and reliability, and the heuristic design principles used made it uncertain that the product improvements that had been achieved on the relatively small SIRE machine could be assured on one that had been scaled up to large heat loads. At the time there was no assurance that A. D. Little would be able to find a material which would solve the thermal distortion problems of its Rcube.

Thus, the Air Force Space Technology Center initiated the Alternative Multi-Stage Cryogenic Refrigerator (AMSCR) program. This program had two primary objectives. The first was to extend the non-wearing refrigerator technology developed by DARPA and NASA to the lower temperatures required by doped silicon LWIR detectors. The second was to respond to the higher heat load requirements then evolving from the Strategic Defense Initiative. It was an eleven-month design study to produce the preliminary design of a flight-quality three-stage cryorefrigerator.

Three contracts were awarded. Garrett Airesearch conducted the preliminary analysis for adding a third expansion turbine to the two-stage turbo-refrigerator technology they had demonstrated on the Mini-HALO program. A. D. Little was selected to explore a variant of their Rcube technology. By plumbing the expansion engines in series rather than parallel, the pressure drop across the low-temperature stage, and hence the bearing load, was reduced to where the clearance could be opened up to 150 millionths of an inch—three times the previous design. The reduced pressure drop required that the flow be increased to maintain the same cooling effect. This was done by increasing the bore, resulting in a much thicker, stiffer piston which was more resistant to distortion. The third contract went to Aerojet for a sorption compressor-driven Joule–Thomson concept.

As a result of these studies, Airesearch selected a third turbine expansion stage for its concept after considering such alternatives as a JT stage or a stored cryogen third stage. Aerojet found that a completely sorption system would require excessive power at present and evolved into a hybrid concept of a single-stage linear Stirling refrigerator for the first stage, a hydrogen sorption-driven JT second stage, and a mechanical compressor-driven JT third stage. A. D. Little refined the design of their low-pressure-drop Rcube concept.

7.6 The PFC program

The AMSCR program served as the de-facto phase zero of a major development effort which was underway in 1989 but has since been cancelled; the Prototype Flight Cooler (PFC) program. The Strategic Defense Initiative Organization decided to make a major commitment to bringing cryogenic refrigerator technology to a point where it was demonstrated to be sufficiently mature for use in an operational space system. To do this, one or more refrigerator designs would be built to spaceflight standards and subjected to performance and life demonstration testing. This would be a technically challenging program, as the envelope of cooling loads and temperatures would be extended well beyond what had been demonstrated to date. A representative design of a typical LWIR sensor using doped silicon detectors was chosen as a 'strawman' requirement upon which to base the design point. The problem statement was relatively close to that of the AMSCR program, with modifications reflecting the greater knowledge of system requirements then available. It called for the machine to provide 2.5 watts of cooling at 8.5 K, 8 watts at 25 K, and 90 watts at 70 K while rejecting to a 285 K sink with a 0.95 success probability for a 50,000-hour mission. It was never intended that this hardware support a specific flight system, or even that the hardware to be built in the PFC program be flown. Rather, it was meant to demonstrate the state of the art at a nominal design point with the understanding that a refrigerator meeting the actual requirements of the prospective spacecraft systems of the 1990s could confidently be scaled from this demonstrated design.

The PFC was conceived as a three-phase program. In Phase I (PFC-I) each contractor was to produce a detailed design of the mechanical portions of the refrigerator to flight standards and of the electronics to laboratory-based standards. Aspects of each design involving unproven technology or considered to be of high risk were to be identified and demonstrated through a critical component test program. This would allow the decision to proceed with fabrication to be made with full confidence that the major technical risks had been resolved, making the later phases a relatively low risk program.

In Phase II (PFC-II) each of the selected contractors was to build two copies of the refrigerator designed in Phase I. In addition, flight-quality electronics were to be designed and fabricated and a design study performed for a cooler with the same temperatures but much larger heat loads. The first hardware item was to be performance tested to its design point, environmentally tested to qualification levels, and then retested at the design point. The second piece of hardware was to be environmentally tested to acceptance levels. The hardware will be life tested in Phase III (PFC-III).

Three contracts were awarded for PFC-I in 1985: A. D. Little, Garrett Airesearch, and Hughes Aircraft. The A. D. Little concept was a direct application of its AMSCR design. The Airesearch concept also was based upon its AMSCR design. It placed all three turbo-expander stages on a single shaft. This achieves a mechanical simplicity and solves the problem of whether foil gas bearings could be made to work with the small high-speed turbine needed to obtain the third-stage temperature.

In recognition that the greatest technical risk was associated with performance of the coldest stage, a subcontract for an optional third-stage design was given to Creare Inc. This was a very small turbine rotating at 600,000 r.p.m., with the shaft suspended on tilting pad hydrodynamic gas bearings. It was designed to be plug compatible with the main Garrett system. As it was optimized to third-stage requirements, its higher specific speed provides higher theoretical efficiency, but at the price of slightly greater system complexity. This high-speed turbine technology was originally sponsored by DARPA as a backup for the Mini-HALO expander in 1982.

Hughes Aircraft was awarded a contract for a new embodiment of the VM cycle that addressed and corrected the shortcomings of earlier designs. It was linearly driven, thus avoiding the side forces, linkages, and bearing associated with the rotary to linear motion. The rubbing seals were replaced by clearance seals, and a passive magnetic bearing system was proposed to prevent contact and wear of the moving expander pistons. Unfortunately, in 1986, Hughes was forced to conclude that the passive magnetic suspension system would not work. They replaced it with an actively controlled magnetic suspension, based on technology developed by NASA/GSFC.

The first phase of this program was completed in March 1987. In August 1987, A. D. Little and Airesearch (now Allied-Signal Aerospace) were selected as contractors for the second phase (PFC-II). The option having all three turbo-expanders on a single shaft was chosen for the Airesearch concept, largely due to cost constraints.

The first hardware components of this program are moving through the fabrication and initial test phases (summer 1989). The program has been modified somewhat in response to perceived changes in user requirements and reductions in the SDIO budget. The work on a system for large heat loads was deleted as sizes of proposed sensor systems were scaled back. Only one complete flight-quality hardware unit is to be built in this phase, and the emphasis upon flight-quality electronics has been reduced. The technical progress at both contractors has been impacted by the effort required to respond to frequent changes in the scope of the program.

Earlier efforts have demonstrated that the technology for producing the required temperatures for heat loads of interest for spacecraft applications does exist. While verifying and improving upon these earlier demonstrations, the PFC-II program is meant to address the issues of producibility, reliability, and the establishment of a well-documented technology base which can be used as the basis for derivative coolers meeting the temperature and load requirements of future user systems. Unfortunately, it appears that to do so in the context of a government contract requires a documentation and control system far more complex than that associated with the technology development programs of earlier refrigerators. The implementation of such a system must be achieved at the expense of the technical effort in an environment of limited resources. As a result, the hardware development and demonstration has not proceeded at the pace one would have anticipated, based upon the amount of prior effort that has gone into both concepts.

7.7 Other programs

The SDIO is sponsoring work on advanced refrigeration technologies which one day may result in machines offering improved reliability or efficiency. Aerojet is continuing work on sorption compression cycles growing out of the concepts developed for the AMSCR program. The simplicity, small volume, and low vibration of this approach is attractive, but the large amount of power required remains a problem. JPL is also being funded to study this technology.

Hughes Aircraft and Aerospace Corporation of America were awarded contracts in 1987 to develop and test machines based upon adiabatic demagnetization. This followed work initiated by DARPA in 1983. The Hughes contract was terminated in 1988. The ACA proof of concept machine produced net refrigeration for a 1.5 K load while rejecting to a 4.5 K sink. A new magnet currently being tested is expected to raise the upper temperature limit to 15 K. This is a potentially important technology, but is still very immature. Its optimum configuration and most appropriate system applications are still in the process of being evolved, and thus this probably should not be thought of as a contender for near-term applications.

The design point of 5 watts at 65 K continues to be of interest. NASA and SDIO have been funding Creare to develop a single-stage reversed Brayton turbo-refrigerator for this design point. A program to fund the development of a flexure supported Stirling for this design point is currently being initiated at the Air Force Space Technology Center.

The above design point appears to be an appropriate temperature but excess capacity for a vast number of potential scientific payloads. Several American aerospace companies have responded to this requirement by initiating the development of flexure-based Stirling units of less than one watt capacity, but there has not yet been a formal government program initiated.

7.8 Summary and trends

The technology of non-contacting, non-wearing seals and bearings has been demonstrated in the laboratory to a degree that long-life, reliable cryorefrigerators can now be designed with high confidence. There are several alternative approaches to reaching 65 K with a single stage, 25 K with two stages, and in the neighbourhood of 10 K with three stages. Unlike some of the devices requiring cryogenic cooling, no further advances in physics are required. The work remaining is straightforward engineering and test. The technologists know that the critical problems have been solved. The next step is to build and test sufficient machines for the potential users to accept that this technology is in fact ready for implementation. This testing will offer many challenges in correcting the inevitable design errors and improving efficiency and reliability, but it will lack the cutting edge glamour of most of the work to date. It will also be very expensive and require a continuing commitment independent of the fortunes of user programs. It may thus be even more difficult to achieve than the progress we have made so far.

The philosophy of achieving life and reliability by designing out mechanical failure modes and reducing the refrigerator to an electronic component may come full circle. As the reliability becomes dominated by electronics, especially for systems designed to operate in enhanced radiation environments, passive self-regulating components such as gas bearings become more attractive than approaches requiring complex electronic sensing and control. As an example of this, if dynamic power systems ever become available for spacecraft, refrigerator reliability could be improved by a direct mechanical drive of the compressor which eliminated the motor invertor circuits.

The Stirling cycle in principle offers the lowest power requirement, except at lower temperatures where regenerator performance degrades. However, most missions in the near future will require that at least two refrigerators be flown. Stirling (and VM) machines will require thermal diodes to switch from the primary to redundant cooler. This technology is not well developed, especially at low temperatures. The reverse Brayton machines have an advantage in this regard as they generate a stream of cold gas that can be directed to wherever needed and, with a long thin connecting tube, the circulation is inherently self-switching. The technology demonstrated in the PFC program covers all warmer temperatures as a subset. The designs can be scaled over a reasonable range of design loads. For loads larger than the PFC design point, it is quite clear that turbo-compression is the only reasonable solution. For smaller loads, positive-displacement systems such as the Rcube become more attractive. Turbine systems have the additional advantage of low vibration, for missions for which that is a consideration. In selecting a refrigerator, the total system penalty including both physical weight and penalty for power and heat rejection must be considered.

Further Reading

Air Force Flight Dynamics Laboratory (1973). Closed Cycle Cryogenic Cooler Technology and Applications. AFFDL-TR-73–149.

Breckenridge, R. W. *et al.* (1986). Preliminary Design of an Alternate Multiple Stage Cryogenic Refrigerator. AFSTC-TR-86–3, Air Force Space Technology Center, New Mexico.

Cleveland, F. R. (1986). Split-Stirling Cycle Cryocooler Off-Design Performance Tests. *Proceedings of International Cryocooler Conference*, Easton, Md.

Leffel, C. S. and Vonbriesen, R. (1981). The APL Satellite Refrigerator Program. QM-80–178, Johns Hopkins University, Laurel, Md.

National Bureau of Standards (1980). Refrigeration for Cryogenic Sensors and Electronic Systems. NBS SP 607, Boulder, Colo.

Robinson, J. W., Giellis, R. T., and Fester, D. A. (1984). Long Term Cryogenic Storage Study. AFRPL TR-83–082, A. F. Rocket Propulsion Lab. Edwards A. F. Base, CA.

Swift, W. (1986). Development of Miniature Turboexpanders for a Long-Life Cryocooler. AFWAL-TR-3067, A. F. Wright Aero. Lab., Ohio.

Wapato, P. G., Norman, R. H., and Barr, K. P. (1982). Application of the Turbo-Refrigerator to Long-Term Cryogenic Storage. SAE Paper 820841, *Twelfth Intersociety Conference on Environmental Systems*.

8

INFORMATION SOURCES ON CRYOGENIC REFRIGERATION

8.1 Introduction

There are many sources of information about cryogenic engineering. We are most familiar with the English literature, much of it of United States origin, and so will deal mostly with that. There are other important sources of cryogenic information, Russian, German, Japanese, British, and French, that should not be overlooked.

The literature can be divided generally into two principal groups: (a) government reports and (b) open literature. It is important to have access to both. Many government reports are never summarized in the open literature. Furthermore, many of the reports, although unclassified, are restricted in circulation to agencies of government and their contractors. Still others are classified at various levels of restriction. Many of the applications of low-capacity cryogenic refrigeration are concerned with military or space systems so the restricted access to many reports is particularly severe.

8.2 Government reports

US Government reports on low-capacity cryogenic refrigeration are, in the main, those of contractors to the government agencies, principally NASA, the Army, the Navy, the Air Force, and the Department of Energy. These are concerned mostly with small or miniature cooling systems for infrared thermal sensing and other electronic applications. The principal agency for US Army cryocooler applications is the Night Vision Laboratory, US Army Research and Development Laboratory, Fort Belvoir, Virginia. The development of rail-gun launchers for high-velocity projectiles has involved considerable cryogenic interest from various Army authorities. The principal agencies for US Air Force cryocooler applications are: the Air Force Flight Dynamics Laboratory, Wright Patterson Air Force Base, Dayton, Ohio, and Philips Laboratories, Kirtland Air Force Base, Albuquerque, New Mexico. The principal agencies for US Navy cryocooler applications are: the Naval Research

Laboratory, Washington, DC, the Naval Engineering Laboratory, Washington, DC, and the David Taylor Model Ship Research and Development Laboratory, Annapolis, Maryland. The principal centres for US National Aeronautics and Space Administration (NASA) interest in cryocoolers include, the Goddard Space Flight Center, Greenbelt, Maryland, and the Ames Research Center, California, with minor interest at the Manned Space Flight Center, Houston, Texas, the Lewis Research Center, Cleveland, Ohio, and the Kennedy Research Center, Cape Canaveral, Florida. The main contractors for cryocooler developments for US Government agencies include: Hughes Aircraft Company, Texas Instruments Inc., Martin Marietta Company, Cryogenic Technology, Inc., Philips Laboratories, R. C. A., Allied Signal, AiResearch Corporation, Aerojet Electrosystems Inc., Magnavox, and General Pneumatics Corporation. There are many other companies involved to a greater or lesser extent in the development and use of cryocoolers. Many of the US national laboratories, non-profit contract research laboratories, and university research laboratories have cryocooler and cryogenic activities. One of the principal centres for cryocooler performance evaluation is the cryogenic engineering group headed by Dr. Ron Ross at the Jet Propulsion Laboratory, California Institute of Technology, Pasadena, California.

United States government reports can sometimes be obtained simply by requesting a copy from the contractors or the agency of government concerned. This only works when the request is made soon after the report is published and spare copies are likely to be on hand. One must, therefore, know the report is due or was recently published. In that fortunate situation, one is probably already on the distribution list and receives the report anyway. However, an unofficial approach is always worth a try, particularly where a report is required urgently, but the chance of a favourable response declines exponentially with elapsed time following publication.

The official sources of United States government reports are:

(1) National Technical Information Service (NTIS)
US Department of Commerce
Springfield, Virginia 22161

(2) Defense Documentation Center (DDC)
Defense Logistics Agency
Cameron Station
Alexandria, Virginia

NTIS handles the distribution of copies of all US government unclassified, unrestricted reports in all fields. DDC supplies copies of reports on defense and security-related matters at all levels of security classification to approved requesters. Only approved 'DDC Users' will likely have success in requesting material from DDC. One becomes a 'DDC User' by working for a US Government agency or government contractor on defense-related matters.

8.2.1 The National Technical Information Service searches

The most important source of cryogenic information and data is contained in the collection of technical reports prepared by contractors to the United States Government. This voluminous and extensive collection covers every conceivable field of human activity. The general collection is held by the National Technical Information Service (NTIS) of Springfield, Virginia. A corresponding collection of classified or restricted reports is held by the Defense Documentation Center, Cameron Island, Virginia.

Anyone from anywhere may request anything from NTIS. The biggest problem is knowing what to request, the proper identification for the document sought, and the way to go about requesting it.

The collection is huge, over 1,000,000 reports with 20,000 added *every day*. A simple request to your friendly, company librarian to 'Get all they have on cryocoolers' would result in the arrival of $17\frac{1}{2}$ tonnes of paper at your office over the next month. One needs to be much more specific.

The way to start is first to consult the NTIS directory of the published collections of abstracts called NTISearches. A collection of abstracts on a particular topic is generated whenever a user requests NTIS to search their collection for entries on, say, miniature Linde–Hampson cryocoolers.

The searches are done using computers instructed by keywords supplied by the user. The computer finds reports with the same keywords and then prints out data about the report including the very important accession number, the title, author's name and affiliation, date, contract number, and the report abstract, a 100–200 word summary of the content of the report.

All these data summaries are compiled into a collection of abstracts with brief, succinct general titles, say 'Linde–Hampson Cryocoolers'. Reference to this collection allows the original requester to study the abstracts, select those he requires, and then request, with all the necessary information to hand, a copy of the original report.

The whole process is fairly lengthy despite valiant efforts by NTIS to speed things up. Special express service at a higher fee can be requested.

A search of the NTIS collection can be commissioned by any user on payment of the appropriate fee, about $200 for up to 500 entries. Unless specified otherwise the search starts with the most recent entries and searches back chronologically until the specified number of entries have been located or the collection exhausted.

Before a new search is commissioned it is advisable to consult the list of searches already completed and now available. The compilations of abstracts that have previously been culled from the NTIS data base are all listed in a directory available on request from NTIS called NTISearches. Copies of these existing compilations are readily available at low cost (about $50). If nothing on the topic can be found the next step is to call the NTIS Sales Desk at 703–487–4650 or the Additional Information Desk at 703–487–4929 to confirm they have nothing published on the topic. Invariably helpful, courteous, and anxious to help, often they can suggest an alternative designation to the one being investigated.

At the present time there are five compilations listed under the title 'Cryogenic Refrigeration' (William E. Read) in NTISearches, specifically:

(1) Volume 1, 1964–1972, NTIS/PS-75/825

(2) Volume 2, 1973–1977, NTIS/PS-78/1261

(3) Volume 3, 1977/1982, NTIS/PB-83–807396

(4) Volume 4, 1982/1983, NTIS/PB-83–807404

(5) Volume 5, 1982/1986, NTIS/PB-86–859204

There are no entries for the specific topic 'Low-Capacity Cryogenic Refrigeration' but many of the entries listed in the above compilations above refer to small, low-capacity, or miniature coolers.

8.2.2 Superintendent of Documents

Another US Government agency responsible for the dissemination of government information is the Superintendent of Documents (SupDocs). This is the sales organization of the US Government Printing Office (GPO). Various reports from government research laboratories are published by the GPO and distributed through SupDocs. They do not handle government contractor reports nor is there any obligation to maintain stocks of particular items, so that frequently one's requests are met with an 'out-of-print' response. NTIS handles the same material as SupDocs plus, of course, the government contractor reports. The difference is that SupDocs material is generally first-edition, original-quality print work whereas NTIS generally supplies xerographic paper copies or microfiche prints of the original report or file. Very occasionally an original copy of the report requested is supplied by NTIS but this only occurs soon after publication and release of the document.

8.2.3 The Cryogenic Data Center

Another US Government source that was very important to those interested in cryogenic information was the Cryogenic Data Center at the laboratories of the National Bureau of Standards (now the National Institute of Science and Technology, NIST), US Department of Commerce, Boulder, Colorado. Over a period of 20 years the Cryogenic Data Center amassed the world's best collection of cryogenic literature. The material was all included in a computerized data base system organized to facilitate subject, author, or chronologic searching. This specialist repository built up over many years was precipitately abandoned in the early 1980s on some stupid and ill-advised bureaucratic whim at the US Department of Commerce. Pirate copies of the computerised data base do exist and may be consulted unofficially, but of course do not contain up-to-date entries. Moreover, much of the printed material, charts, and monographs available have now been distributed or otherwise disposed of but hopeful enquiries to former members of the Cryogenic Data Center can still result in the provision of helpful material. A good compilation of the NBS cryogenic charts and tables of properties data is reproduced in Barron (1984) and Timmerhaus and Flynn (1990). The charts are too small to be of practical use but happily the print quality is so good that enlargements for general use can be readily obtained by xerographic or photographic means. Large-scale transparencies that can be used for dye-line prints are particularly useful for routine repetitive work involving use of the charts.

8.2.4 Conference proceedings

From time to time agencies of the US Government organize specialist conferences dealing with cryocoolers and matters relating thereto.

Infra-Red Information Symposia (IRIS) are held annually (about May/June) at various centres in the United States. The symposia are a joint service-classified meeting organized by the Office of Naval Research devoted to military applications of infrared radiation. The proceedings of the symposia are edited by the Infrared Information and Analysis Center, Environmental Research Institute of Michigan, P. O. Box 8618, Ann Arbor, Michigan.

Various other government-sponsored symposia have been held from time to time. Although not specific to cryocoolers the proceedings do contain significant contributions of interest. The Office of Naval Research and the National Aeronautics and Space Administration are notable organizers of space and other application-oriented workshops with much of cryogenic interest. A Space Cryogenics Workshop is organized usually to more or less coincide with the biennial Cryogenic Engineering Conference. Unfortunately there appears to be no formal organizational or publications structure for the proceedings of the workshop.

8.3 The open literature

8.3.1 Introduction

Publishers with a long and consistent interest in cryogenics include:

(1) Oxford University Press, Oxford, UK: The Monographs on Cryogenics Series.

(2) Plenum Publishing Corporation New York:

- (a) International Monographs on Cryogenics and other books on cryogenics;
- (b) *Advances in Cryogenic Engineering* (Proceedings of the Cryogenic Engineering Conference and the International Cryogenic Materials Conference, now about 40 volumes in total);
- (c) *Journal of Low Temperature Physics*.

(3) Butterworths Heinemann, Linacre House, Jordan Hill, Oxford, UK:

- (a) *Cryogenics*, a scientific periodic journal;
- (b) Proceedings of the International Cryogenic Engineering Conference;
- (c) *Journal of Refrigeration*, a publication of the International Institute of Refrigeration, Paris.

8.3.2 Books

There are several excellent books on cryogenic engineering and cryogenic systems. The best currently available and the foundation for any cryogenic-related collection includes books by Barron (1984) and Hands (1986) and more recently Timmerhaus and Flynn (1990). Happily the classic work on cryogenic engineering (Scott 1988) has recently been reprinted and is still a most valuable reference to the field. A volume specializing in the topic of miniature cryogenic refrigerators for cryogenic sensors and cold electronics was published by Walker (1989). The earlier two-volume work on cryocoolers by Walker (1983) is the most extensive collection of material assembled on the topic of cryocoolers generally and includes a surprisingly voluminous bibliography. This was assembled with the aid of the Cryogenic Data Center at the National Bureau of Standards. Walker's book published in 1983 was actually completed in 1979 so even the most recent material contained is now at least 15 years old.

Of course the 40 or so volumes in the serial publication *Advances in Cryogenic Engineering* (ACE) published by Plenum, New York, represent the most extensive and coherent collection of the cryogenic literature. These volumes are the proceedings of the bi-annual Cryogenic Engineering Conference and the International Cryogenic Materials Conference.

8.3.3 International Cryocooler Conference

Throughout the last decade a substantial measure of scientific and engineering attention has been focused on the topic of low-capacity refrigeration for cryogenic sensors, for superconducting devices, and for aerospace applications. A specialist conference, the International Cryocooler Conference has become established on a biennial cycle in the United States. Its focus is primarily on low-capacity cryocoolers rather than large-scale plants.

The first of these conferences was held at the US Department of Commerce National Bureau of Standards Laboratory at Boulder, Colorado, in the Fall of 1977. A small group (40 or so) specialists in the field was invited to severally address the problems of small cryocoolers. The proceedings of this meeting were published (Zimmerman *et al.* 1978) and sufficient interest was aroused to justify a second meeting organized in 1980 (Zimmerman *et al.* 1981).

This second meeting was open and anyone from anywhere could attend. It has come to be called the First International Cryocooler Conference. Among *aficionadoes* the initial, private, meeting is often referred to as the Zeroth International Cryocooler Conference.

The third meeting, the Second International Cryocooler Conference, was held in December 1982 at the National Aeronautics and Space Administration Goddard Space Flight Center, Greenbelt, Maryland (Gasser *et al.* 1983). The fourth meeting, the Third International Cryocooler Conference, returned to the NBS Laboratories, at Boulder, Colorado in the Fall of 1984 (Radebaugh *et al.* 1985). The next meeting was held at Easton, Maryland, in Fall 1986 organized by the David Taylor Naval Ship Research and Development Center, Annapolis, Maryland (Green *et al.* 1987). The sixth meeting was held in August 1988 at the Naval Postgraduate School, Monterey, California (Lindqvist 1989). The seventh meeting took place in Plymouth, Massachusetts, in November organized by the David Taylor Naval Ship Research and Development Center (Green and Knox, 1990). The Seventh International Cryocooler Conference was held in Santa Fe, New Mexico, under the auspices of the US Air Force, Kirtland Air Force Base (Ludwigson, 1993). The Eighth meeting is scheduled to occur in Vail, Colorado in 1994 under the chairmanship of Dr. Ron Ross, Jet Propulsion Laboratory, Pasadena, California.

A compilation of the abstracts of papers presented at the first six conferences, the Gifford Memorial Volume, was assembled (Walker *et al.* 1990) and is a useful starting point for new topics. It is anticipated that a second edition will be published to include abstracts of the first ten conferences.

The international cryocooler conferences were organized to be synchronized more or less with the biennial Applied Superconductivity Conference (who deal with all aspects of superconducting systems except cryocoolers) and held either just before or just after the larger superconductivity conference. For recent meetings, however, this congruence has not been so evident and is not likely to be continued in the future.

Involvement of the military has resulted unfortunately in the creation of a 'closed session', the Inter-Agency meetings, at the Cryocooler Conference with attendance limited, and the publication of special, restricted-distribution proceedings.

The collected proceedings of the eight cryocooler meetings together with a compilation of the abstracts constitutes the best summary of contemporary research and development interest in low-capacity cryogenic refrigerators.

Reading these proceedings will, however, quickly reveal that for many topics there is considerable reticence to reveal much detail of principal substance. From the academic aspect this is to be deplored although it is perhaps inevitable given both the 'security classified' and 'commercially confidential' pressures.

It is unfortunate, too, the companies with the largest cryocooler activities choose to contribute only rarely to the open sessions of the cryocooler conference. They probably present a paper or two at the Inter Agency meeting and feel they have then made their contribution.

8.3.4 Other cryogenic conferences

Other conferences that regularly include papers of significant interest about cryogenic refrigeration are:

(1) the Cryogenic Engineering Conference (CEC);

(2) the International Cryogenic Engineering Conference (ICEC);

(3) the International Cryogenic Materials Conference (ICMC).

The Cryogenic Engineering Conference The Cryogenic Engineering Conference occurs in the United States, presently on a biennial basis. It is organized by the Cryogenic Engineering Conference Board, an *ad-hoc* group of interested individuals including representatives of the co-sponsors, such as the International Institute of Refrigeration (IIR) based in Paris and the principal engineering societies.

The International Cryogenic Materials Conference Another conference, the International Cryogenic Materials Conference (ICMC) is held simultaneously with the CEC and at the same location. The proceedings of both the CEC and the ICMC are published as *Advances in Cryogenic Engineering* (ACE) by Plenum, New York. Volumes in the ACE series now include 40 or more sizeable books. Together they constitute the greatest collection of material on the topic of cryogenic engineering dating back to the foundation meeting in 1954, the start of US effort in the space race.

The first 25 or so volumes were edited in meticulous and painstaking detail by Klaus Timmerhaus, Associate Dean of Engineering at the University of Colorado. Both Timmerhaus and Plenum deserve an accolade for their extraordinary investment of time and attention to detail in this massive contribution to cryogenic engineering. Happily the editorial traditions established by Timmerhaus have been sustained by his successors in maintaining the uniform character of the series.

Volume 20 of ACE contains a comprehensive index of the first 20 volumes and is the first source consulted by this author when a new topic of interest in cryogenic engineering arises. Hopefully, with ACE now into Volume 40, the publishers will see fit to commission the preparation of another comprehensive index covering the second 20 volumes, perhaps integrating it with the present index in Volume 20.

The International Cryogenic Engineering Conference The International Cryogenic Engineering Conference is held every two years, at a location other than the United States, in the years when the Cryogenic Engineering Conference is not being held. Unfortunately, of course, this coincides (or conflicts?) with the International Cryocooler Conferences, also held in the years when the CEC is not taking place. In 1992 for example the fourteenth International Cryogenic Engineering Conference was held, in July, at Kiev, Ukraine, and was followed in November by the Seventh International Cryocooler Conference. In 1994, the fifteenth ICEC is scheduled to be held, followed shortly thereafter by the eighth International Cryocooler Conference.

The proceedings of the International Cryogenic Engineering Conference are published in England by Butterworths Heinemann, Linacre House, Jordan Hill, Oxford UK.

8.3.5 Cryogenics

Butterworths also publish 'Cryogenics', the only international scholarly journal devoted entirely to low-temperature research and development in all its aspects. The journal started in 1960 under the guidance of Professor K. Mendelssohn of the University of Oxford Clarendon Laboratory. Cryogenics publishes a broad range of papers on research topics, general topical and book reviews, reports of conferences, and notices of interest to the cryogenics community.

The Journal of Refrigeration, of the International Institute of Refrigeration in Paris, France, is also published by Butterworths. This journal is concerned with refrigeration over a broader temperature range but often contains papers of interest to the cryogenic community.

Several books of interest are also published by Butterworths including the *Cryogenics Handbook* by Beverly Law (1976) now sadly dated and in urgent need of an update.

8.3.6 International Institute of Refrigeration

The International Institute of Refrigeration (IIR), located in Paris, is the oldest international organization devoted to refrigeration technology. Presently the main meetings of the Institute are held every four years and are called the International Congress of Refrigeration. The proceedings are published (in three or four volumes) under the title of *Progress in Refrigeration Science and Technology*.

The IIR is organized in ten interest groups, called Commissions, as follows:

(1) cryophysics and cryoengineering;

(2) heat and mass transfer;

(3) refrigerating machinery;

(4) refrigeration of perishable produce;

(5) cold-storage facilities;

(6) air conditioning;

(7) refrigerated land transport;

(8) refrigerated sea transport;

(9) applications of refrigeration to chemical, civil, and industrial engineering;

(10) cryobiology and freeze drying.

In addition to the regular quadrennial Congress meetings, the various Commissions meet either singly or in combinations having mutual interests to consider specific topics or a range of topics. For example, one meeting of Commission 1 in 1970 addressed itself to measurement of temperature, He^3/He^4 refrigeration, heat transfer to liquid gases, superconductivity, and experimental techniques. In London in 1969, one meeting of Commission 1 was concerned with liquid natural gas; simultaneously, but at a separate meeting in London, the topic was low-temperature and electric power. The proceedings of all these meetings are published by the IIR as a supplement or annex of the Bulletin, a regular bimonthly publication of about 250 pages including review papers, original studies, information on current research in refrigeration, information on forthcoming meetings of interest, and the program abstracts of papers of the next Congress.

In addition to all these, the IIR has many other important publications including three *Bibliographic Guides to Refrigeration* for the periods 1953–60, 1961–4, and 1965–8, an international dictionary of refrigeration, and many different charts and tables of thermodynamic properties.

The veritable plethora of activities, interests, and publications with many variations due to the input of local characteristics makes consultation of their publications somewhat difficult. To obtain a list of publications or further information, application may made to:

Institut International du Froid
(International Institute of Refrigeration)
177 boulevard Malesherbes
F 75017 Paris
France

The Institute is better known in Europe than in the United States. However, there are sufficient members of the Institute in the United States to maintain liaison with the specialist cryogenic organizations.

8.3.7 Low-temperature physics

The cryogenics field has been a playground of the physicist for a hundred years, several times as long as it has been a topic of substantial engineering interest. Therefore, it is important not to overlook the physics sections of the library. Many choice items of great interest to cryogenic engineers will be found there. For example, Daunt has given an excellent and very comprehensive treatment of 136 pages, a book in itself, in the *Handbuch der Physik*, Vol. XIV, on 'The Production of Low Temperatures down to Hydrogen Temperature'.

Then there are the proceedings of the Conference of Low Temperature Physics. Meetings are held at three-year intervals in different parts of the world, with the proceedings of the conference produced by the local organizing group. The proceedings appear to be published by a variety of organizations and publishers. Consultation of the various physics journals is recommended for current information.

8.3.8 The American Society of Heating, Refrigerating, and Air Conditioning Engineers

The American Society of Heating, Refrigerating, and Air Conditioning Engineers (ASHRAE) is a society based in the United States but with a strong international presence and is concerned with all aspects of air handling and processing. Although they have little specific interest in the field of cryogenics they do publish much material with valuable cryogenic information, notably the ASHRAE *Handbook of Fundamentals*. This volume contains the most complete and readily available collection of thermodynamic data and charts for refrigerants including many of the cryogenic fluids.

8.4 Newsletters and periodicals

8.4.1 Low Temperature News

This is the newsletter of the British Cryogenics Council. This extremely interesting periodic publication is edited and produced on behalf of the British Cryogenics Council and is sponsored by the Institutions of Chemical Engineers, Civil Engineers, Electrical Engieneres, Gas Engineers, Mechanical Engineers, and the Institutes of Measurement and Control Metals, Physics, Refrigeration, and the Society for Low Temperature Biology.

The newsletter provides a forum for the exchange of news and information in the field of cryogenics. It contains announcements of forthcoming events, press releases, articles, and other material relating to cryogenics. The present editor (1993) is Mr. J. S. Harris, 5 Foxhill Rd., Burton Joyce, Nottingham, NG14 5DB, UK, Telephone/Fax 0602 312896.

8.4.2 Cold Facts

This is a publication similar to the above but tends to deal with developments in North America whereas Low Temperature News focuses more on British and European events. Cold Facts is published by the Cryogenic Society of America Inc., and the Records Secretary is Mrs. Laurie Huget, 1033 South Blvd., Suite 13, Oak Park, Illinois 60302, USA, Telephone 708–383–6220, Fax 708–383–9337.

8.4.3 Cryo Gas International ISSN 1052–0139

Previously known as the Cryogenic Information Report—ISSN 0011–2259, this is published 10 times a year by J. R. Campbell at Associates Inc., 5 Militia Drive, Lexington, MA 02173, USA, Telephone 617–862–0264, Fax 617–863–9411. Ms. Kay Deans is the editor. This is an exceptionally valuable periodic report on all phases of cryogenic interest with details of US patents, foreign patents, company literature, research and development, gas applications, markets, technology, updates (contemporary developments), and so on. Citations include the contents of:

(1) *Gas and Chemical Reports*: latest reports from Japan concerning the gas business;

(2) *Journal of Low Temperature Physics* published by Plenum, 233 Spring St., New York, NY 1003;

(3) *Cryogenics* published by Butterworth Heinemann, Linacre House, Jordan Hill, Oxford, OX2 8DP, UK;

(4) *Membrane and Separation Technology News* published by BCC Inc., 25 Van Zant St., Suite 13, Norworld, CT 06855, Telephone 203–853–4266.

(5) *The Hydrogen Letters* published by Peter Hoffman, 4104 Jefferson St., Hyatsville, MD 20781, Telephone 301–779–1561, Fax 301–927–6345.

(6) *Liquified Natural Gas Express* published by Zeus Development Corporation, 3827 Villanova St., Houston, TX 77005–3639, Telephone 713–664–1621.

8.4.4 Stirling Machine World

Stirling Machine World (SMW) is a specialist periodical published four times a year and addressed specifically to Stirling machine specialists. It contains a miscellany of news, views, tidbits of information, notices of forthcoming meetings, abstracts of conference papers, review of new books, news of interesting new Stirling machines, and the like. It is published by Brad Ross of 1823 Hummingbird Court, Richland, Washington, USA 99352–9542.

8.4.5 House journals

Two house journals have, over the years, contributed very substantially to the cryogenic engineering literature, particularly with regard to cryogenic cooling systems.

One of these two journals is the *Philips Technical Review*. Since the first publication, by J. W. L. Kohler in 1954, of two articles about the new Philips Stirling cooling engines, there have been many other papers dealing with subsequent developments and improvements. The *Philips Technical Review* is an excellent journal devoted to research topics and new product development in the Philips groups of companies. Subscriptions or single copies of the *Review* may be obtained on application to:

N.V. Uitgeversmaatschappij Centrex
(Centrex Publishing Co.)
N.W. Emmasingel 9
P.O. Box 76
Eindhoven, Netherlands

Another journal of exceptional relevance and interest to the field of cryocoolers is the *Linde Reports in Science and Technology*. This monthly publication contains six to ten articles describing recent Linde developments and plant construction in the field of cryogenic gas processing and chemical engineering. An index to the articles or copies of the reports may be obtained on application to:

Pressestelle der Linde Aktiengesellschaft
Wiesbaden, Hildastrasse 2–10
West Germany

8.4.6 Database computer searches

Computer based information services are offered by most good, technical libraries using governmental or commercial facilities to search data bases of government reports, conference proceedings, scholarly journals, the proceedings and transactions of the technical societies, and the periodicals. With most topics of interest it is necessary to be very specific in order to avoid being 'snowed' in paper. This is done using carefully selected keywords.

For example we might use the single keyword *refrigerator* and request a search of all the available data bases for a period of the past 50 years. Provided your librarian took you seriously this widespread, unspecific, request would likely result in enough paper to keep you occupied until retirement.

A better choice of keywords would be: *refrigerator, cryogenic, Linde–Hampson, Joule–Thomson, miniature, low-capacity, superconducting, semiconductor, electronics*. A second search might be requested with the same keywords except for *Stirling, mechanical* substituted for *Linde–Hampson, Joule–Thomson*.

Should the response to a highly specific request result in a disappointingly small amount of references one can always return with a less tightly structured request. The process can be repeated as often as desired to generate whatever volume of response is considered necessary or desirable.

8.4.7 Liquid Natural Gas Conferences

The development of intercontinental transportation of liquid natural gas in the 1960s and later was a major application of cryogenics technology. Principal interest is, of course, focused on very large production, storage, and transportation facilities Nevertheless there is much interest to be found in the proceedings of the quadrennial Liquid Natural Gas Conferences.

These conferences are organized by the International Gas Union, a body representing all aspects of natural gas technology throughout the world. The most recent Liquid Natural Gas Conference, the thirteenth, was held in Kuala Lumpur in September 1991.

8.4.8 Institute of Gas Technology

Details of the LNG conference and copies of the past proceedings may be obtained from the Publications Department of the Institute of Gas Technology (IGT), 3400 State St., Chicago 60649, Illinois. IGT publishes much other material of cryogenic interest. The complete list of publications may be obtained on application.

A feature of life in the late 1980s and early 1990s is the unprecedented growth of interest in the use of natural gas as a vehicle fuel. Organizations such as the Natural Gas Vehicle Coalition, the American Gas Association, the Gas Research Institute, the US Department of Energy, and various state energy and environmental agencies mount conferences, short courses, and specialist meetings about natural gas as a vehicle fuel. The Institute of Gas Technology, Library and Informations Division is wonderfully well informed and equipped to provide the specialized information and details of references available on payment of a small user fee.

8.4.9 Manufacturers' literature

Finally, an excellent source of valuable and detailed information is, of course, the publications of the cryocooler manufacturers. These include product data sheets, technical memoranda, product handbooks, reference manuals, and maintenance handbooks. Often the manufacturers have training manuals that are invaluable to newcomers to the field.

8.4.10 Short courses

The Continuing Education Division of the University of California, Los Angeles (Engineering Director, Dr William Goodin, Telephone 310–825–3344, Fax 310–206–2815) has provided annually a one-week course on cryogenic engineering (run by Kropschot, Timmerhaus, Edeskuty, and Flynn) for over 30 years. The textbook for the course is Timmerhaus and Flynn (1990).

A similar one-week short course on Cryocooler has been provided for the past 12 years by Radebaugh and Walker, *et al.* The new text for this course is this present volume.

Another short course of interest provided occasionally by the Continuing Education Division of the American Society of Mechanical Engineers and the American Society of Chemical Engineers is Cryogenic Fundamentals by Randall Barron, author of the foundation book *Cryogenic Systems* (Barron 1984).

All the above courses are available for presentation elsewhere in variable format at company or government agency locations, universities, research institutes, private organizations, and the like anywhere in the world. Interested parties should contact directly the various authors listed above.

8.4.11 Project study allocation

With any new project the difficulty is knowing when to stop searching and start looking in detail at the papers of greatest relevant to the topic of interest. The accumulated literature of the world is now so vast and, through the co-operative efforts of the libraries and information services, so readily available one could simply fill every day reading material relevant to the topic of interest (and of course all the other interesting topics that arise serendipitously).

In recognizing this and avoiding the temptation to read too much there is also the danger of reading too little. Most engineers and scientists will, in fact, acknowledge they do not spend enough time in consulting the established record of past endeavours before starting on a new project. Reinventing the wheel is a common failing and many can attest to the wisdom of the old adage 'He who fails to study history is doomed to relive it.' Project managers and administrators are always anxious for the project engineers to 'stop poncing about and start doing something useful'. However, decisions made on the basis of inadequate or inappropriate information or data sometimes lead the project into a swamp from which it is costly in terms of time, effort, and money to later escape, particularly after metal has been cut and substantial time and money spent.

It is difficult to say how long or how much effort should be invested in literature research at the start of a new project. Less than 10 per cent of the time and effort available is probably too little, more than 30 per cent is probably too much. In this we include instructional ventures such as continuing education courses at specialist centres and attendance at the specialist conferences and exhibitions.

8.5 Conclusion

There is an extensive literature on cryocoolers although few specific books on the topic. The guide to the literature included here should help most people get started. There are many other books and sources of information that could legitimately have been included; our sincere apologies are offered here to all who are affronted by our apparent neglect of their work.

References

Barron, R. (1984). *Cryogenic Systems*. 2nd edn, Monographs in Cryogenics 3, Oxford University Press.

Daunt, J. G. (1956). The Production of Low Temperature down to Hydrogen Temperatures. In *Handbuch der Physik*, BD IV, Springer-Verlag, Berlin.

Gasser, M. *et al*. (1983). *Proceedings of the Second International Cryocooler Conference*, NASA Goddard Space Flight Center, Greenbelt, Md.

Green, G. and Knox, M. (1991). *Proceedings of the Sixth International Cryocooler Conference*, David Taylor Naval Ship R. and D. Center, Annapolis, Md.

Green, G., Patton, G., and Knox, M. (1987). *Proceedings of the Fourth International Cryocooler Conference*, David Taylor Naval Ship R. and D. Center, Annapolis, Md.

Hands, B. (1986). *Cryogenic Engineering*. Academic Press, London, UK.

Law, B. (1976). *Directory of Cryogenics*. Butterworths Scientific Publishers, Guildford, UK.

Lindqvist, P. (1989). *Proceedings of the Fifth International Cryocooler Conference*, Naval Postgraduate School, Monterey, Ca., Aug.

Ludwigson, J. (1992). *Proceedings of the Seventh International Cryocooler Conference*, Kirtland AFB, Albuquerque, N.M.

Radebaugh, R., Louie, B., and McCarthy, S. (1985). *Proceedings of the Third International Cryocooler Conference*, National Bureau of Standards Special Publication 698, US Government Printing Office, Washington, D.C., May.

Ross, R. (1995). *Proceedings of the Eighth International Cryocooler Conference*, Jet Propulsion Laboratory, California Institute of Technology, Pasadena, Ca.

Scott, R. B. (1988). *Cryogenic Engineering* (reprint). Met-Chem Research Inc., Boulder, Colo.

Timmerhaus, K. and Flynn, M. (1990). *Cryogenic Process Engineering*. Plenum, New York.

Walker, G. (1983). *Cryocoolers*. 2 vols., Plenum, New York.

Walker, G. (1989). *Miniature Refrigerators for Cryogenic Sensors and Cold Electronics*. Monographs on Cryogenics, Oxford University Press.

Walker, G. and Bingham, E. R. (1990). Gifford Memorial Volume: Abstracts of Papers presented at six International Cryocooler Conferences, 1977, 1980, 1982, 1984, 1986, 1988, Department of Mechanical Engineering, University of Calgary, Alberta, Canada.

Zimmerman, J. *et al.* (1978). *Proceedings of the Zeroth International Cryocooler Conference*, National Bureau of Standards, Boulder, Colo.

Zimmerman, J. *et al.* (1981). *Proceedings of the First International Cryocooler Conference*, National Bureau of Standards, Boulder, Colo.

APPENDIX

Address list of contacts with interests in miniature cryocoolers

Reproduced from '*Survey of the State of the Art of Miniature Cryocoolers for Superconducting Devices*, J. L. Smith, G. Robinson, and Y. Iwasa, Massachusetts Institute of Technology, Cryogenic Engineering Laboratory, Cambridge, Mass. 02139, January 1984.

Canada

Quantum Technology Corporation
2528 Yale St.
Vancouver, BC VSK 1B9
Calvin Winter, 604–682–6488

France

A. B. G. Semca
3 Villa Thorenton
3 Paris

CENG/SBT
38041 Grenoble Cedex
J. Verdier, (76) 974111

L'Air Liquide
38360 Sassenage,
Isere
A. Faure, (76) 268131

Germany

A. E. G. Telefunken
Theresien Str. 2, Heilbrun 7100
H. Korf

Leybold-Heraeus Gmbh
Bonner Strasse 504
D 5000 Koln 41
R. Heisig, 0221–3701

Linde A. G.
D 8023 Hoellriegelskreuth
Munich
F. Vosswinkel, 089–7273

Messer Griesheim Gmbh
Homberger Strasse 12
D 4000 Dusseldorf

Siemens A. G.
Paul Gossen Strasse 100
D 8520 Erlangen
G. Bogner

Holland

NV Philips USFA B. V.
NL 5600, ND Eindhoven
N. Baerselman, 040–784167

Israel

Ricor Ltd.
18960 En Harod 1 hud

Italy

Officene Galileo
Localita Tomerello 50013
Camp Bisenzo Florence
D. Soldaini, 055–89501

Japan

Aisin Seiki Company Ltd.
1 Asahi-machi 2-chome
Kariya city, Aichi-ku 448
T. Tani, 0566–24–8337

Cryogenic Systems
253–5 Yamanouchi
Kamakura 247
Y. Ishizaki, 0467–25–3825

Electrical Laboratory
1–14 Umezono Sakuramura
Niiharugun Ibaraki 305
S. Tomiyama, 0298–54–5262
N. Tamada, 0298–54–5262

Hitachi Kasado Works
794 Higashitoyoi Kudamatsu City
Yamaguchi-ken
S. Saito, 0833–41–0123

Hitachi Works
4–6 Kanda Surugadai
Chiyoda-ku, Tokyo
M. Hachiya, 03–258–1111

Kobe Steel Ltd.
Asada Research Laboratory
53–3 Gomo-aza-maruyama
Nada-ku Kobe
J. Tsukada, 078–251–1551

Mitsubishi Electric Corporation
Central Research Laboratory
1–1 Tsukaguchi-Honmachi 8-chome
Amagasaki,
Hyogo-ken 661
M. Iwamoto, 06–491–8021

Nihon University
Kanda Surugadai
Chiyoda-ku, Tokyo
K. Yasukochi, 03–293–3251
Y. Matsubara, 0474–66–1111 x 6480

Osaka Oxygen Industries Ltd.
12–4-2-chome Utajima
Nishiyodogawa-ku, Osaka
M. Yanai, 06–472–1181

Sumitomo Heavy Industries, Ltd.
1 Kander Mitoshiro-cho
Chiyoda-ku, Tokyo 101
K. Kikuchi, 03–296–5631

Suzuki Shokan
3–1 Kojimachi
Chiyoda-ku, Tokyo
Y. Hiresaki, 03–265–5211

Tohoku University
Department of Physics
Sendai 980
T. Ohtsuka, H. Fujita, 0222–402–6231

Toshiba Corporation
Energy Science and Technology
Laboratory
4–1 Ukishima-cho
Kawasaki 210
H. Oguiwara, W4–277–3111

Ulvac Cryogenics, Inc.
Hagisono 2500 Chigasaki-City
Kanagawa-ken 253
H. Sibata, 0467–85–6522
H. Hayakawa, W67–58–4367

Switzerland

Sulzer Bros. Ltd.
CH 8401, Winterthur
H. Quack, 41–52–813728

United Kingdom

British Aerospace Dynamics Division
Manor Rd.
Hatfield, Devonshire
0707–262300

Cryogenic Consultants Ltd.
Metrostore Blvd.,
231 The Vale
London W32QS
J. Good, 7436049

Horstman Defence Systems Ltd.
Cocksbrook Rd.
Bath, Avon BA13EX
0225–23111

Hymatic Engineering Company Ltd.
Burnt Meadow Rd.
North Moons Moat
Redditch B989HJ
R. McLellan, 52764931

International Research and
Development Company Ltd.
Fossway
Newcastle-Upon Tyne NE624D
A. D. Appleton, 0632–650451

Oxford Instrument Company
Osney Mead
Oxford OX20DX
R. B. Inwood, 0865–241456

Oxford University Clarendon
Laboratory
Oxford, OX13PU
G. Davey, 0865–59988

Thor Cryogenics Ltd.
Henley Rd.
Berinsfield,
Oxford
P. J. Hatton, 0865–340601

United States

Aerojet Electrosystems Company
1100 W. Hollyvale St.
Asuza, CA 91202
R. J. Williams, 213–812–1864

The Aerospace Corporation
300 North Aviation Blvd.
El Segundo, CA 90009
M. Donabedian, 213–648–6315

AiResearch Manufacturing Company
2525 W. 190th St.
Torrance, CA 90507
O. Buchmann, 213–512–3393

Air Products and Chemicals Inc.
1833 Voltec St.
Allentown, PA 18015
R. Longsworth, 215–481–3708

Ball Aerospace Systems Division
Boulder Industrial Park
Boulder, CO 80306
R. P. Reinker, 303–441–4000

Balzers High Vacuum Corporation
Hudson, NH 03060
B. Hamilton, 603–889–6888

Beech Aircraft Corporation
Boulder, CO 80301
W. A. Edgington, 303–443–1650

Carlisle Cryotronics
114 Sunset Rd.
Carlisle, MA 01741
D. Sarcia, 617–369–0785

Creare Inc.
Hanover, NH 03755
W. Swift, 603–643–3800

Cryomech Inc.
1630 Erie Blvd.
East Syracuse, NY 13210
P. Gifford, 315–475–9692

Cryosystems Inc.
190 Heatherdown Dr.
Westerville, OH 43081
D. Swartz, 614–882–2796

CTI-Cryogenics
266 Second Ave.
Waltham,
MA 02254
P. Kerney, 617–890–9400

CVI Inc.
Columbus, OH 43216
T. Hickey, 614–876–7381

Energy Research and Generation Inc.
Lowell and 57th St.
Oakland, CA 94608
G. Benson, 415–658–9785

Ford Aerospace and Communications Corporation
Ford Rd.
Newport Beach, CA 92660
R. Grabow, 714–720–7108

General Pneumatics Corporation
440 Washington St.
Orange, NJ 07050
S. Zylstra, 201–678–6555

H. R. Textron
25200 West Rye Canyon Rd.
Valencia, CA 91355
L. Drazin, 805–253–5215

Hughes Aircraft Company
2000 East El Segundo Blvd.
El Segundo, CA 90245
J. J. Orth, 213–616–0724

Hughes Aircraft Company
3100 West Lomita Blvd.
Torrence, CA 90505
J. Johnson, 213–517–6196

Kryovac Scientific
Columbia Drive
Amherst,
NH 03031
C. Lamb, 603–883–3232

Lakeshore Cryotronics
64 E. Walnut St.
Westerville, OH 43018
W. Pierce, 614–891–2243

A. D. Little Inc.
Acorn Park
Cambridge, MA 02140
W. Breckenridge, 617–864–5770

Lockheed Palo Alto Research Laboratory
3251 Hanover St.
Palo Alto, CA 94304
T. Nast

MMR Technologies Inc.
1400 Stierlin Rd.
Mountain View, CA 94043
W. Little, 415–962–9620

Magnavox Government and Industrial Electronics Company
46 Industrial Ave.
Mahwah, NJ 07430
D. Lehrfeld, 201–529–1700

Martin Marietta Denver Aerospace
12250 S. Hwy 75
Denver, CO 80201
C. Class, 303–977–8739

Mechanical Technology Inc.
986 Albany-Shakey Rd.
Latham, NY 12110
R. Ackerman, 518–785–2211

Metal Bellows Corporation
1075 Providence Hwy
Sharon, MA 02067
C. Carswell, 617–668–3050

Oxford-Airco
600 Milik St.
Carteret, NJ 07009
G. Taylor, 201–541–1300

Philips Laboratories
Scarborough Rd.
Briarcliff Manor,
NY 10510
A. Daniels, 914–945–6257

Quantum Design Inc.
11404 Sorrento Valley Rd.
San Diego,
CA 92121
R. Sager, 619–457–0248

Rockwell International
Anaheim, CA

Santa Barbara Research
75 Coromar Dr.
Santa Barbara,
CA 93017
C. Olaina, 805–968–3511

S. H. E. Corporation
4174 Sorrento Valley Blvd.
San Diego,
CA 02121
D. Paulson, 619–453–6300

Texas Instruments
Dallas, TX
R. Dix, 214–995–2011

TRW Inc.
One Space Park
Redondo Beach, CA 90278
A. H. Silver

Varian
645 Clyde Ave.
Mountain View, CA 94043
415–964–2790

US Army Night Vision Laboratory
Fort Belvoir, VA 22060
S. Horn, 203–664–1345

Air Force
Wright Patterson AFB, OH
C. Oberly, 513–257–1110

Jet Propulsion Laboratory
4800 Oak Grove Dr.
Pasadena, CA 91109
E. Tward, 213–354–6581

Los Alamos Scientific Laboratory
Los Alamos, NM
John Barklay, 505–667–6431

National Bureau of Standards
325 Broadway
Boulder, CO 80303
J. Zimmerman, 303–497–3901
R. Radebaugh, 303–497–3710

NASA Goddard Space Flight Center
Greenbelt, MD 20771
M. Gasser

NASA Ames Research Center
Moffett Field, CA 94035
P. Kittel, 415–965–6530

NAME INDEX

INSTITUTION INDEX

SUBJECT INDEX